Developmental Systems

S. Węgrzyn J.-C. Gille P. Vidal

Developmental Systems

At the Crossroads of System Theory, Computer Science, and Genetic Engineering

With 74 Figures

Springer-Verlag
New York Berlin Heidelberg
London Paris Tokyo Hong Kong

Stefan Węgrzyn
Institute of Theoretical and Applied Computer Science of the Polish Academy of Sciences
44100 Gliwice
Poland

Jean-Charles Gille
Faculty of Science and Engineering
Laval University
Quebec G1K 7P4
Canada

Pierre Vidal
Centre d'automatique
Université des sciences et techniques de Lille
59655 Villeneuve d'Ascq
France

Printed on acid-free paper.

Typeset by Compélec, Inc., Quebec, Canada
Printed and bound by Edwards Brothers, Inc., Ann Arbor, Michigan.
Printed in the United States of America.

9 8 7 6 5 4 3 2 1

ISBN 0-387-97340-0 Springer-Verlag New York Berlin Heidelberg
ISBN 3-540-97340-0 Springer-Verlag Berlin Heidelberg New York

PREFACE

Many facts were at the origin of the present monograph. The first is the beauty of maple leaves in Quebec forests in Fall. It raised the question: how does nature create and reproduce such beautiful patterns? The second was the reading of A. Lindenmayer's works on *L* systems. Finally came the discovery of "the secrets of DNA" together with many stimulating exchanges with biologists.

Looking at such facts from the viewpoint of recursive numerical systems led to devise a simple model based on six elementary operations organized in a generating word, the analog of the program of a computer and of the genetic code of DNA in the cells of a living organism.

It turned out that such a model, despite its simplicity, can account for a great number of properties of living organisms, e.g. their hierarchical structure, their ability to regenerate after a trauma, the possibility of cloning, their sensitivity to mutation, their growth, decay and reproduction. The model lends itself to analysis: the knowledge of the generating word makes it possible to predict the structure of the successive developmental stages of the system; and to synthesis: a specific type of structure can be obtained by systematically constructing a generating word that produces it.

In fact the model here proposed is coherent with the fundamental assumptions of cellular biology and in particular with recent discoveries concerning DNA, which in the light of our model behaves like a very elaborate generating word.

This monograph represents the present state of our research.

It is the authors' ambition that this work will help system engineers become acquainted with these problems and will suggest some hypotheses to biologists.

* *
*

It is a pleasure for the authors to acknowledge their gratitude to Professors Jean-Louis Lavoie and Gérald Lemieux and to Dr. Gene Bourgeau of Laval University in Québec, who helped them clarify some points of mathematics and of biology respectively, and to Drs. Adam Mrózek and Ryszard Winiarczyk of the Institute of Theoretical and Applied Computer Science of the Polish Academy of Sciences in Gliwice, who computed the table in Chapter 3.

We are deeply indebted to Professors H.P. Baltes of University of Alberta in Edmonton, Guy Jumarie of Université du Québec in Montréal and Andrew P. Sage of Mason University in Fairfax, Virginia for their support and encouragement.

We are grateful to the editors of the magazines in which the greater part of the material here presented was first published, with special reference to the *International Journal of Systems Science* and to *Kosmos*.

The manuscript was typed by Mrs. Hélène Michel and the drawings drawn by Mr. Gaston Guay at Laval University. The type-setting was done by Compélec, Inc. in Québec. To all we express our gratitude. We also thank Mr. Rosaire Giguère who was kind enough to help for the tedious task of proof-reading.

The research carried out was made possible by a grant from the Natural Sciences and Engineering Research Council of Canada.

CONTENTS

Preface

Chapter 1 Proem: The model

1.1 General 1
1.2 Introductory examples 1
1.3 General definition 12
References 19

Chapter 2 Synchronous developmental systems without feedback

2.1 Generating word and system development: analysis and synthesis 21
2.2 Structural properties 34
2.3 Quantitative growth 48
References 60

Chapter 3 Synchronous developmental systems with feedback

3.1 General 63
3.2 Analysis and synthesis 65
3.3 Structural properties 70
3.4 Quantitative growth 77
References 89

Chapter 4 Developmental systems with operating system

4.1 General 91
4.2 Sensitivity to external context 93
4.3 Sensitivity to internal context 95
4.4 Dwindling and regrowth 103
4.5 Multilevel development 110
4.6 On global operating systems 117
4.7 Conclusion 119
References 119

Other Books by the Same Authors

Chapter 1

PROEM: THE MODEL

1.1 GENERAL

A developmental system, in the most general sense, is a *set of ordered, oriented elements* — the *components* — which changes in time.

More particularly, systems will be considered the structure of which changes at *discrete intervals of time* as a consequence of elementary *operations* acting on the elements.

By analogy with living organisms, which consist of cells and develop in time according to certain patterns, it will be assumed that an element ("cell") of a developmental system can undergo differentiation or multiplication (giving rise to not more than two cells) and always derives from a preexisting cell of the system.

1.2 INTRODUCTORY EXAMPLES

1.2.1 Example A: one-dimensional developmental system

Suppose that an initial oriented cell a gives rise at each new developmental stage to two cells a and b (a being of the same, b of a different type) and the b cells then remain unaltered.

The first operation will be called *linear generation* and denoted

$$L(a) = ba \qquad \text{or} \qquad a \rightarrow ba$$

which means that an a cell when acted upon by an L operation gives rise to two cells a and b, situated close to each other, a following b in the direction defined by the oriented a mother cell (Fig. 1-1).

The fact that the b cell remains itself will be expressed as a *stagnation* operation and denoted

$$S(b) = b \qquad \text{or} \qquad b \rightarrow b$$

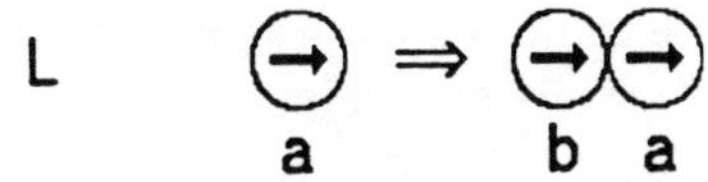

Fig. 1-1. Linear generation.

We shall say that the *generating word* of the system is

$$GW_A = LS$$

and write in expanded form

$$\begin{array}{cc} L & a \rightarrow ba \\ | & \\ S & b \rightarrow b \end{array} \qquad (1\text{-}1)$$

The developmental system will be called

$$DS_A = \frac{\mathrm{L\,S}}{a,b} \qquad (1\text{-}2)$$

with the initial condition: one a cell.

The result is a developmental system DS_A of cells of the a and b categories which changes in discrete time (denoted $k = 0,1,2,\ldots$).

On the basis of (1-1) the successive developmental stages are easily obtained:

$$\begin{array}{ll} k = 0 & DS_A(0) = a \\ k = 1 & DS_A(1) = ba \\ k = 2 & DS_A(2) = bba \\ \ldots & \ldots \\ k = \mathrm{n} & DS_A(\mathrm{n}) = \underbrace{bb\ldots b}_{n}a \end{array}$$

or, in recursive notation:

$$\left.\begin{array}{l} DS_A(0) = a \\ DS_A(k) = b\,DS_A(k-1) \qquad \text{for} \quad \mathrm{k} > 0 \end{array}\right\} \qquad (1\text{-}3)$$

At its k-th stage the system consists of one a cell and of k b cells. If the growth is assumed to take place along a constant direction, the system has the shape of a one-dimensional filamentous string (see Figure 1-2).

The evolution of the system in time can be quantitatively characterized by the number of cells of each category at each stage of its development

$$a(k) = 1 \qquad b(k) = k \qquad (1\text{-}4a)$$

or, in a more compact form, by the row matrix

$$\mathbf{Y}(k) = [a(k) \quad b(k)] = [1 \quad k] \qquad (1\text{-}4b)$$

and its growth rate can be evaluated on the basis of the total number of cells ("volume" or "size" of the system) (Fig. 1-3)

$$V(k) = a(k) + b(k) = k + 1 \qquad (1\text{-}5)$$

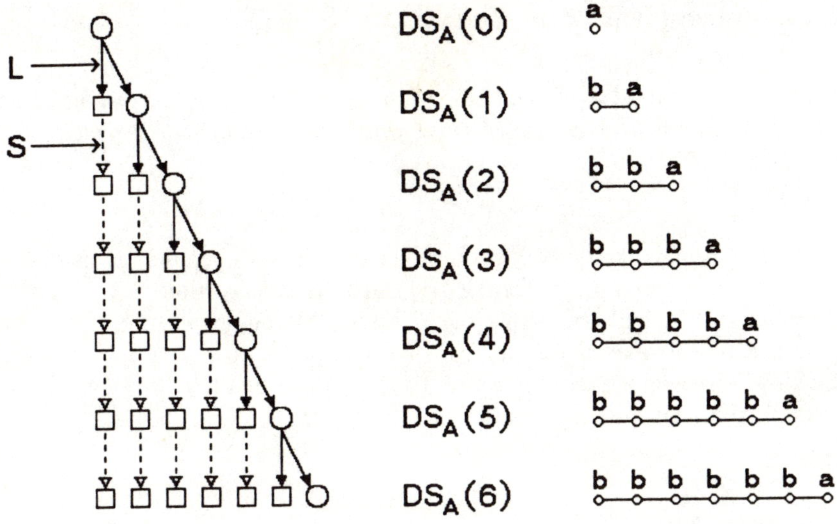

Fig. 1-2. Developmental stages $k = 0$ to $k = 6$ of system $DS_A = \frac{LS}{a,b}$.

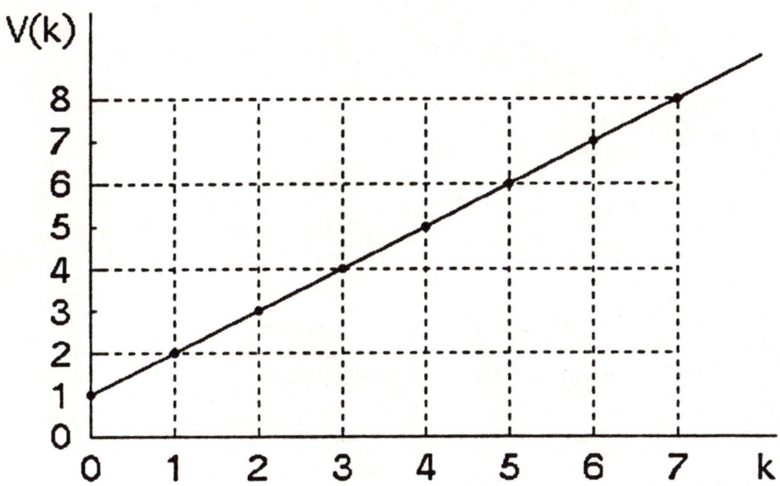

Fig. 1.3. Growth of developmental system DS_A.

1.2.2 Example B: two-dimensional developmental system

Now suppose that the b cell of Example A no longer remains unaltered, but divides at each stage into two other cells, according to an operation which will be called *bifurcation with change of direction* and denoted

$$C_{90}(b) = c(d) \qquad \text{or} \qquad b \rightarrow c(d)$$

This means that a b cell when subjected to a C_{90} operation gives rise to two new cells c and d, the former being still in the direction of the original oriented b cell and the latter being at an 90° angle with it. Here the parenthesis indicates a 90° change of direction for the d cell (Fig. 1-4). Suppose furthermore that the c and d cells remain unaltered throughout the evolution of the system.

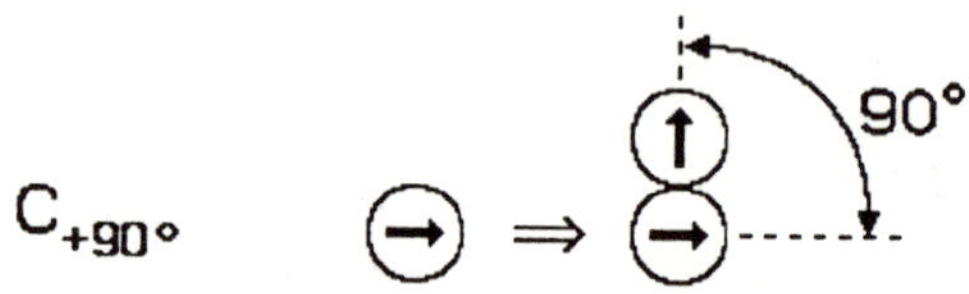

Fig. 1.4. Bifurcation with change of direction of 90°.

The generating word is now

$$GW_B = L\,C_{90}\,S\,S$$

or, in expanded form:

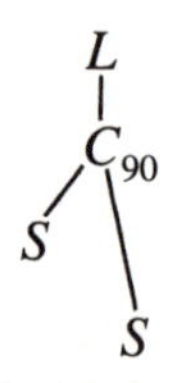

$$a \rightarrow ba$$
$$b \rightarrow c(d) \qquad (1\text{-}6)$$
$$c \rightarrow c$$
$$d \rightarrow d$$

and the developmental system will be called

$$DS_B = \frac{L\,C_{90}\,S\,S}{a,b,c,d} \qquad (1\text{-}7)$$

with one a cell as initial condition.

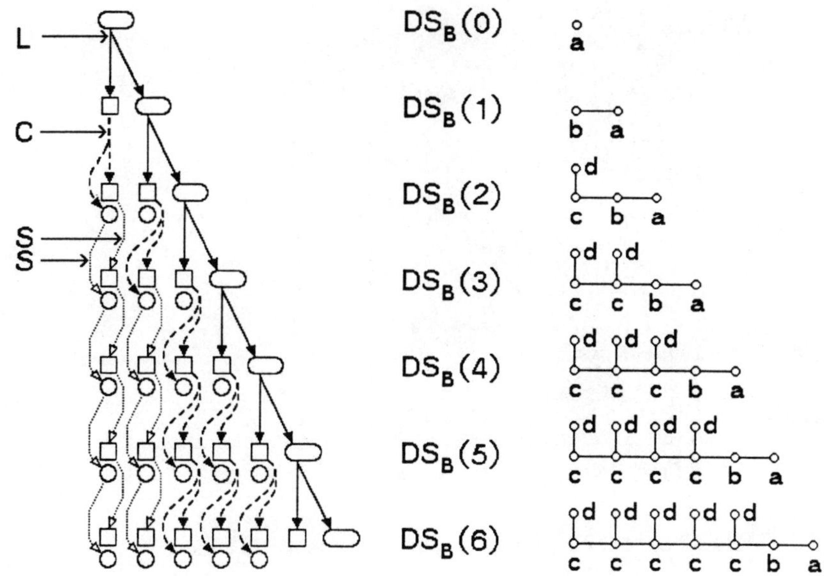

Fig. 1.5. Developmental stages $k = 0$ to $k = 6$ of system $DS_B = \dfrac{L\,C_{90}\,S\,S}{a,b,c,d}$.

The successive developmental stages are easily obtained on the basis of (1-6):

$$k = 0 \qquad DS_B(0) = a$$
$$k = 1 \qquad DS_B(1) = ba$$
$$k = 2 \qquad DS_B(2) = c(d)ba$$
$$k = 3 \qquad DS_B(3) = c(d)c(d)ba$$
$$\ldots \qquad \ldots \qquad \ldots$$

or, in recursive notation:

$$\left.\begin{aligned} &DS_B(0) = a\,, \quad DS_B(1) = \mathrm{ba} \\ &DS_B(k) = c(d)DS_B(k-1) \qquad \text{for} \quad \mathrm{k} > 1 \end{aligned}\right\} \tag{1-8}$$

As a result a two-dimensional pattern is obtained (Fig. 1-5).

The number of the cells at the k-th developmental stage is for $k > 1$

$$a(k) = 1 \qquad b(k) = 1 \qquad \mathrm{c(k) = k - 1} \qquad \mathrm{d(k) = k - 1} \tag{1-9a}$$

$$\mathbf{Y}(k) = [1 \quad 1 \quad \mathrm{k} - 1 \quad \mathrm{k} - 1] \tag{1-9b}$$

whence the size (Fig. 1-6)

$$V(k) = a(k) + b(k) + c(k) + d(k) = 2k \tag{1-10}$$

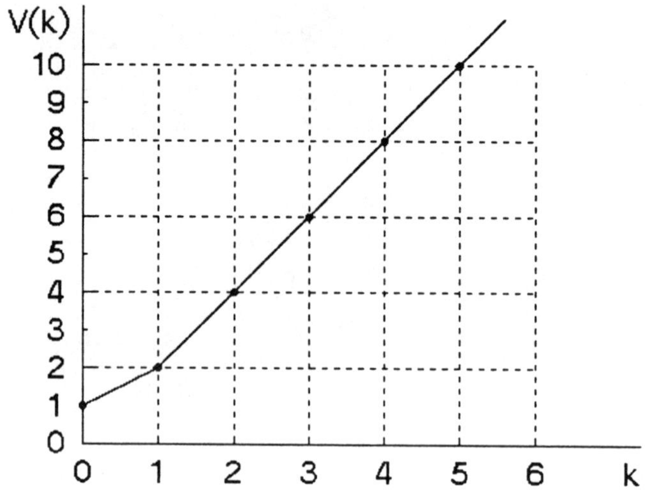

Fig. 1.6. Growth of developmental systems DS_B and DS_C.

1.2.3 Example C: another two-dimensional system

Now suppose that the direction in which the *d* cells develop is not at 90° any more, but makes an angle of 45° with the main development direction, alternately on one side and on the other. In order to distinguish from Example B, a $\pm 45°$ subscript will be added to *C*

$$DS_C = \frac{L\,C_{\pm 45}\,S\,S}{a,b,c,d} \tag{1-11}$$

Equation (1-8) remains valid but the meaning of the parentheses has changed, since they now denote a change of direction of $+45°$ and $-45°$ alternately.

The two-dimensional pattern obtained (Fig. 1-7) has the shape of a tree with branches of unit length. A distinction can be made between the "stagnant" cells, acted upon by *S* operations (*c* and *d*) and the cells which undergo division (*a* and *b*) and are therefore responsible for the system growth. In terms of cellular physiology (Hall, Flowers and Roberts, 1974, pp. 5-8) the latter constitute the *meristem* of the tree-shaped pattern obtained. (In Figure 1-2 the meristem consists of the *a* cell, in Figure 1-5 of the *a* and *b* cells.)

The quantitative evaluation of the cellular composition and of the growth rate [eqs. (1-9) and (1-10)] are of course the same as for Example B.

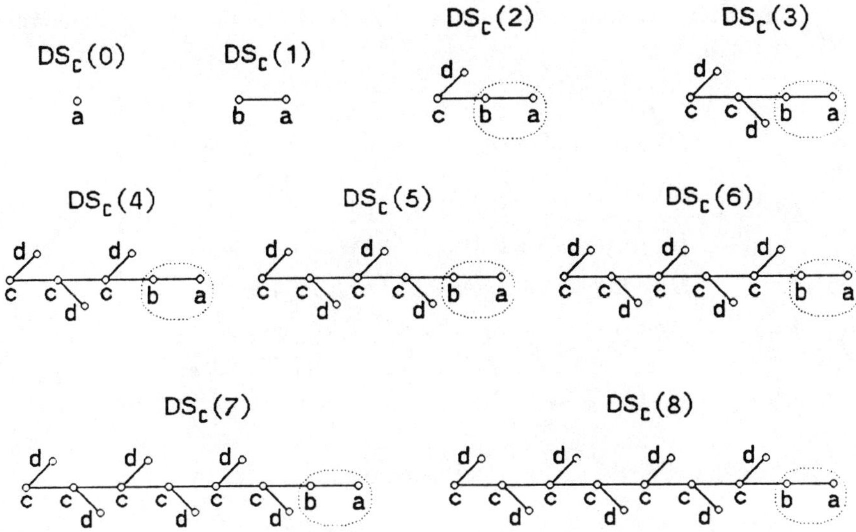

Fig. 1.7. Developmental stages $k = 0$ to $k = 6$ of system $DS_C = \dfrac{L\,C_{\pm 45}\,S\,S}{a,b,c,d}$. (Meristem is outlined.)

1.2.4 Example D: developmental system with feedback

In the previous examples each cell had an antecedent with the exception of cells of the first category (a). A new feature will now be presented: the presence of *feedback* in the sequence of the operations, in the sense that a cell of the first category is regenerated by cells corresponding to the last letter of the generating word.

Consider the developmental system

$$DS_{D1} = \frac{L\,C\,S\,L^F}{a,b,c,d}$$

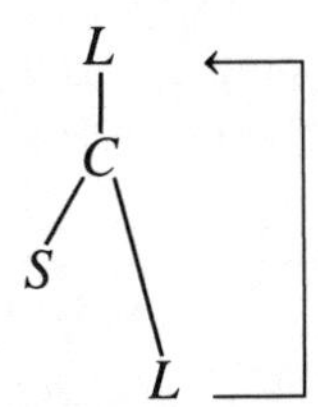

$$a \rightarrow ba \qquad b \rightarrow c(d) \qquad c \rightarrow c \qquad d \rightarrow ad \tag{1-12}$$

where C should be understood as $C_{\pm 45}$ (as in § 1.2.3) and the F superscript indicates the presence of feedback from the second L of the generating word.

The first developmental stages, starting from the initial condition $DS_{D1}(0) = a$, are:

$$\left.\begin{array}{lll} k=1 & DS_{D1}(1) = & ba \\ k=2 & DS_{D1}(2) = & c(d)ba \\ k=3 & DS_{D1}(3) = & c(ad)c(d)ba \\ k=4 & DS_{D1}(4) = & c(baad)c(ad)c(d)ba \\ k=5 & DS_{D1}(5) = & c(c(d)babaad)c(baad)c(ad)c(d)ba \\ k=6 & DS_{D1}(6) = & c(c(ad)c(d)adc(d)babaad) \\ & & c(c(d)babaad)c(baad)c(ad)c(d)ba \end{array}\right\} \quad (1\text{-}13)$$

The result has the shape of a tree with branches and subbranches (Fig. 1-8).

Quantitatively the cellular composition

$$\mathbf{Y}(k) = [a(k) \quad b(k) \quad c(k) \quad d(k)]$$

is given by

$$\begin{array}{llllll} \mathbf{Y}(0) = [\ 1 & 0 & 0 & 0] & \qquad V(0) = & 1 \\ \mathbf{Y}(1) = [\ 1 & 1 & 0 & 0] & \qquad V(1) = & 2 \\ \mathbf{Y}(2) = [\ 1 & 1 & 1 & 1] & \qquad V(2) = & 4 \\ \mathbf{Y}(3) = [\ 2 & 1 & 2 & 2] & \qquad V(3) = & 7 \\ \mathbf{Y}(4) = [\ 4 & 2 & 3 & 3] & \qquad V(4) = & 12 \\ \mathbf{Y}(5) = [\ 7 & 4 & 5 & 5] & \qquad V(5) = & 21 \\ \mathbf{Y}(6) = [12 & 6 & 9 & 10] & \qquad V(6) = & 37 \end{array}$$

The growth (Fig. 1-9) is faster than in Examples A, B and C, but no explicit expression for $V(k)$ is apparent.

In the next example the feedback consists in the fact that a cell of the first category is regenerated by cells corresponding to a non-last letter of the generating word.

Consider the developmental system

$$DS_{D2} = \frac{L\,C^F\,S}{a,b,c}$$

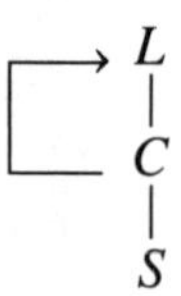

$$a \to ba \qquad b \to c(a) \qquad c \to c \qquad (1\text{-}14)$$

where C should be understood as $C_{\pm 45}$ and the F superscript indicates the presence of feedback from the C operation.

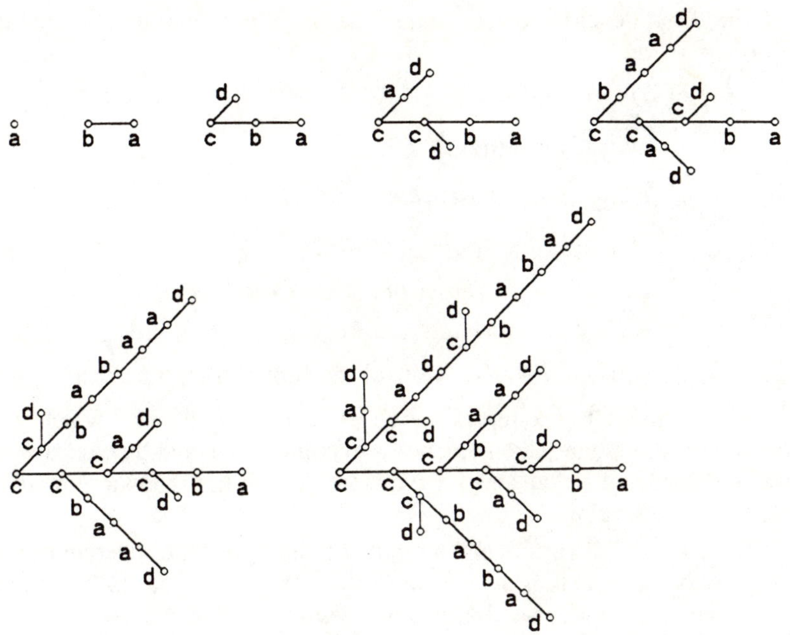

Fig. 1.8. Developmental stages $k = 0$ to $k = 6$ of system $DS_{D1} = \dfrac{LCSL^F}{a,b,c,d}$.

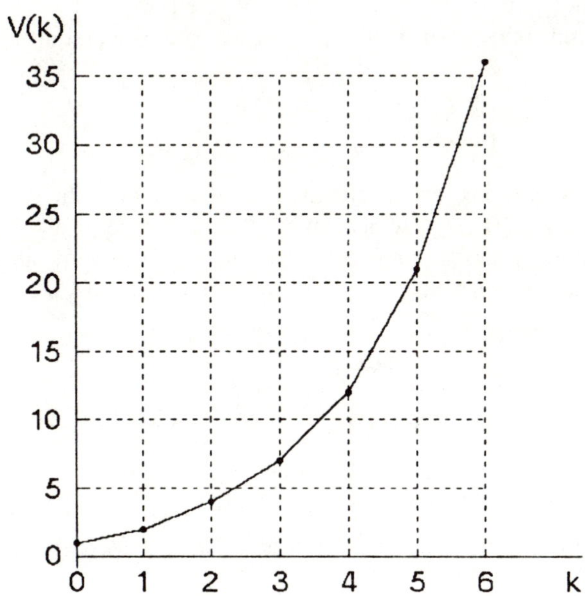

Fig. 1.9. Growth of developmental system DS_{D1}.

The first developmental stages, starting from the initial condition $DS_{D2}(0) = a$, are:

$$DS_{D2}(1) = ba$$
$$DS_{D2}(2) = c(a)ba$$
$$DS_{D2}(3) = c(ba)c(a)ba$$
$$DS_{D2}(4) = c(c(a)ba)c(ba)c(a)ba$$
$$DS_{D2}(5) = c(c(ba)c(a)ba)c(c(a)ba)c(ba)c(a)ba$$

1.2.5 Example E: developmental system with operating system

All the previous examples were developmental systems which grow indefinitely. A system which reaches a maximum size and then decays will now be considered. It consists of Example A with the following modifications of the operations.

a) Instead of always being submitted to an L $(a \to ba)$ operation, the a cell generates ba so long as no more than six b cells are present in the system, but it dies afterwards; in other words, the L operation is replaced by

$$O_aL \quad \begin{cases} a \to ba & n_a \leqslant 6 \\ a \to b & n_a > 6 \end{cases} \tag{1-15}$$

b) Instead of indefinitely remaining itself $(b \to b)$, a b cell remains itself only ten times, then it disappears, i.e. the S operation is replaced by

$$O_bS \quad \begin{cases} b \to b & n_b \leqslant 10 \\ b \to - & n_b > 10 \end{cases} \tag{1-16}$$

What operations are performed is thus under the control of the *operating system* (O_a, O_b), which makes the system respectively sensitive to internal context (number of b cells present in the system) and to external context (discrete time elapsed). The system will be denoted

$$DS_E = \frac{(O_aL)\,(O_bS)}{a,b} \tag{1-17}$$

The first six developmental stages are identical to those of Example A [eq. (1-3)]:

$$DS_E(0) = a \qquad DS_E(1) = ba \qquad DS_E(2) = bba \qquad DS_E(3) = bbba$$
$$DS_E(4) = bbbba \qquad DS_E(5) = bbbbba \qquad DS_E(6) = bbbbbba$$

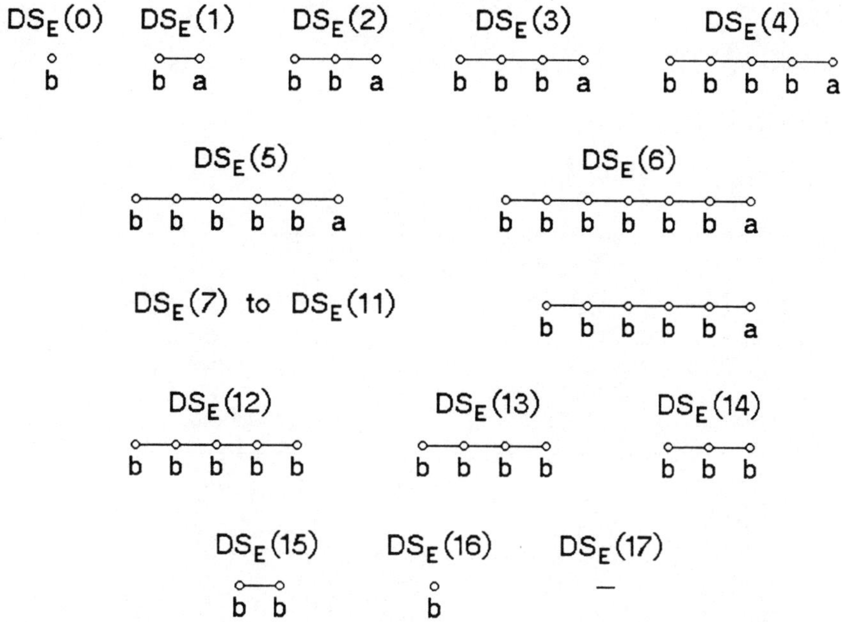

Fig. 1.10. Developmental stages $k = 0$ to $k = 17$ of system $DS_E = \dfrac{(O_aL)\,(O_bS)}{a,b}$.

Then, as a consequence of the presence of six b cells, the a cell disappears, the b cells remaining stagnant:

$$DS_E(7) = bbbbbbb = DS_E(8) = DS_E(9) = DS_E(10) = DS_E(11)$$

At the $k = 12$ developmental stage the $O_b(S)$ operation acting on the last b cell becomes $b \rightarrow -$, whence

$$DS_E(12) = bbbbbb$$

and similarly

$$DS_E(13) = bbbbb \qquad DS_E(14) = bbbb \qquad DS_E(15) = bbb$$

$$DS_E(16) = bb \qquad DS_E(17) = b \qquad DS_E(18) = DS_E(19) = \ldots = -$$

These results are shown in Figures 1-10 and 1-11.

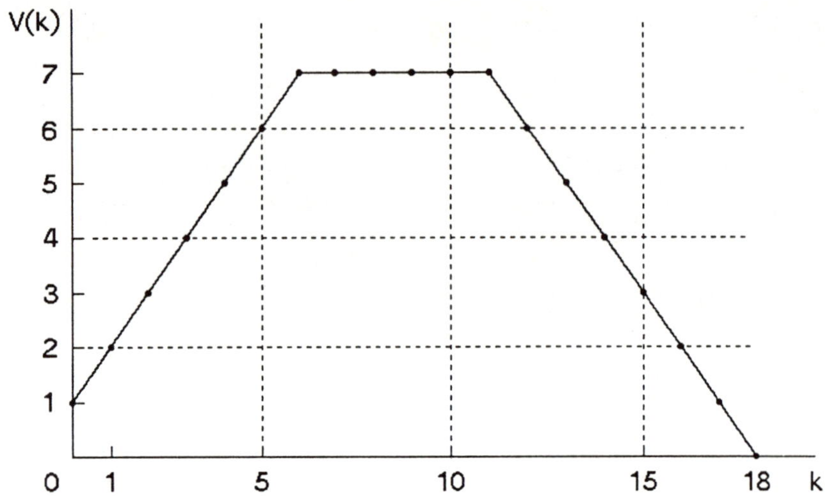

Fig. 1-11. Growth and decay of developmental system DS_E.

1.3 GENERAL DEFINITION

1.3.1 General

A *developmental system DS* consists of:

a) an ordered set Z of *elements* ("cells") of n categories

$$Z = \{a_1, a_2, \dots a_n\}$$

b) a sequence of n elementary operations $A_1, A_2, \dots A_n$ (the *generating word GW*) acting on the cells at discrete instants of time k

$$GW = A_1 A_2 \dots A_n = \underset{i=1}{\overset{n}{\mathrm{SEQ}}} A_i \tag{1-18}$$

c) an initial condition ($k = 0$) which is usually one cell of the first category $DS(0) = a_1$

We shall note:

$$DS = \frac{GW}{Z} = \frac{\underset{i=1}{\overset{n}{\mathrm{SEQ}}} A_i}{a_1, a_2, \dots, a_n} \tag{1-19}$$

1.3.2 The elementary operations

Six types of A_i operations will be considered.

1) The simplest operation consists of a cell remaining unaltered. We shall call it *stagnation* and write.

$$S(a_i) = a_i \qquad \text{or} \qquad a_i \rightarrow a_i \qquad a_i \in Z \qquad (1\text{-}20)$$

2) The next simple operation consists of a cell being differentiated into another. We shall call it *transformation* and write

$$T(a_i) = a_j \qquad \text{or} \qquad a_i \rightarrow a_j \qquad a_i,\ a_j \in Z \qquad (1\text{-}21)$$

3) A cell may divide, giving rise to a new cell. We shall call this *linear generation* and write

$$L(a_i) = a_j a_i \qquad \text{or} \qquad a_i \rightarrow a_j a_i \qquad a_i,\ a_j \in Z \qquad (1\text{-}22)$$

If the generation of the a_j cell occurs with a change of direction, we shall speak of *rotative generation* and write

$$R_\alpha(a_i) = a_j(a_i) \qquad \text{or} \qquad a_i \rightarrow a_j(a_i) \qquad a_i,\ a_j \in Z \qquad (1\text{-}23)$$

where the parenthesis denotes the change of direction (Fig. 1-12).

It may be useful to specify of how many degrees the change of direction consists and whether it always occurs in the same direction (giving rise to a spiralling pattern) or whether the direction alternates at each k (giving rise to a sort of zig-zag pattern). To do this, one may add a subscript and write

$$R_\alpha \qquad \text{or} \qquad R_{\pm\alpha}$$

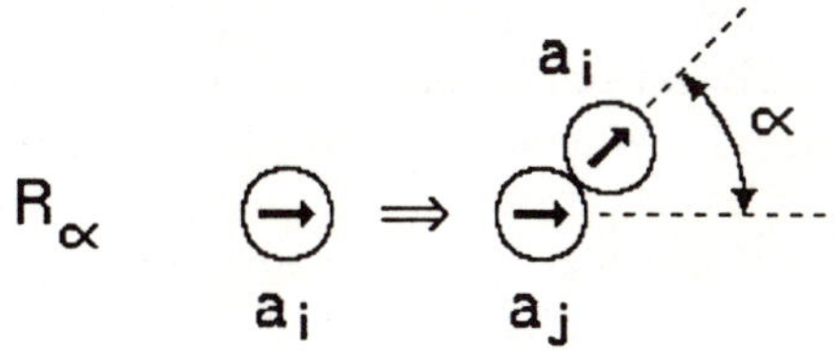

Fig. 1-12. Rotative generation R_α.

4) Finally a cell may *divide* into two cells of different categories:

$$B(a_i) = a_j a_k \qquad \text{or} \qquad a_i \rightarrow a_j a_k \qquad a_i,\ a_j,\ a_k \in Z \quad (1\text{-}24)$$

Such an operation will be called *bifurcation*. The a_j cell is oriented in the direction of the oriented a_i mother cell. If the a_k cell is oriented at an

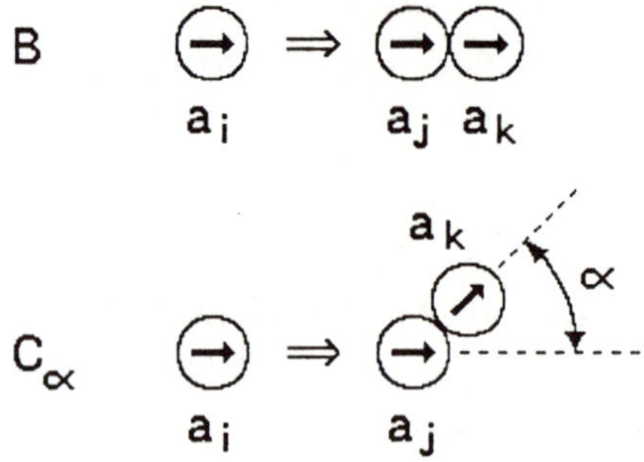

Fig. 1-13. Bifurcation without and with change of direction α.

angle α with respect to the said direction (see Figure 1-13), we shall speak of *bifurcation with change of direction* and write

$$C_\alpha(a_i) = a_j\,(a_k) \qquad \text{or} \qquad a_i \to a_j\,(a_k) \qquad a_i, a_j, a_k \in Z \qquad (1\text{-}25)$$

where again the parentheses denote the change of direction. Like for the R operation, whether the sign of the change in direction is constant or is alternately + and − can be specified by writing

$$C_\alpha \qquad \text{or} \qquad C_{\pm\alpha}$$

In all that follows C without a subscript should be understood as $C_{\pm 45}$.

The cells which undergo L, R, B and C operations constitute the *meristem* of the system. If the said operations would cease being performed the development of the system would also stop.

These six operations (see Figure 1-14) are consistent with axioms of biology: each cell derives from a preexisting one and can at most generate two cells at a time. They have been chosen because it turns out that combining them enables one to account for a great number of properties of two-dimensional developmental systems which are analogous to properties of living organisms (see Chapters 2, 3 and 4).

1.3.3 The generating word

A) General

The sequence of the operations $A_1, A_2, \ldots, A_n$ acting on the n elements $a_1, a_2, \ldots, a_n$ is the *generating word* of the developmental system.

SYMBOL	OPERATION	
	Algebra	Configuration
L	a → ba	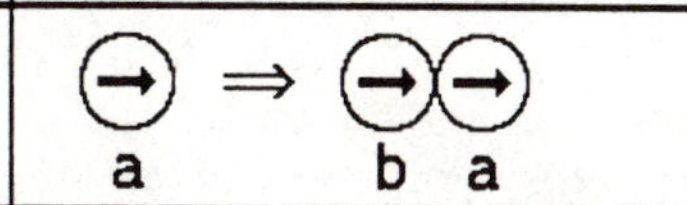
R_α	a → b(a)	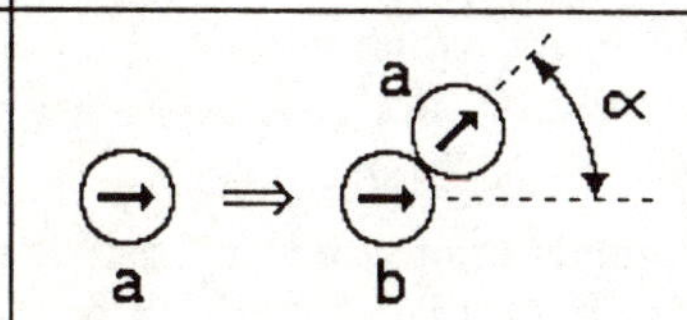
B	a → bc	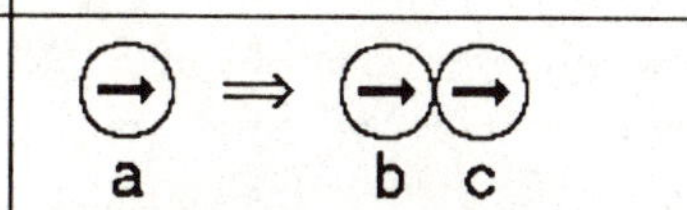
C_α	a → b(c)	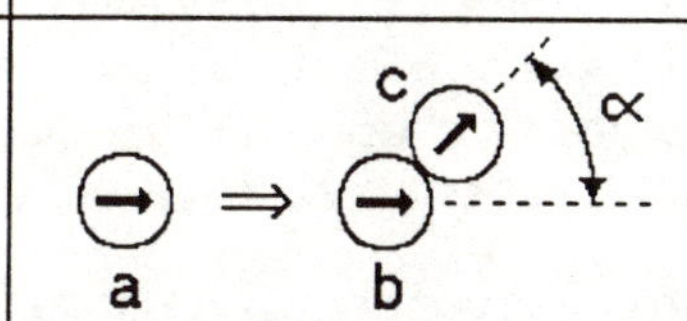
T	a → b	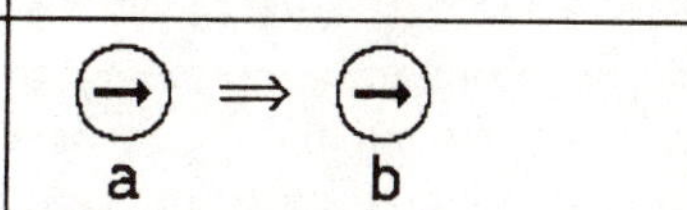
S	a → a	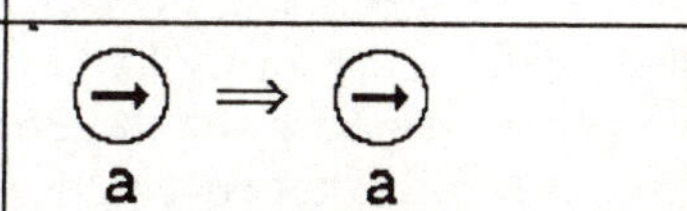

Fig. 1-14. The elementary operations.

If the symbols A_i of the operations are considered as the letters of a *language*, the operation sequence is a word in that language. A "correct" word should fulfill the following conditions:

$$\left.\begin{array}{l} \text{a) if } A_i(a_i) = a_j, \text{ then } j \geqslant i \\ \text{b) if } A_i(a_i) = a_j a_k \text{ or } a_j(a_k) \text{ or } a_j a_i \text{ or } a_j(a_i) \\ \qquad \text{then } j \geqslant i \qquad k \geqslant i \qquad j \neq k \end{array}\right\} \quad (1\text{-}26)$$

The operations should be performed after one another, the first operation acting on the initial a_1 cell. This amounts to "reading" the word from left towards right.

If the conditions (1-26) are fulfilled the generating word will be said to be of the *linear*, or *rectilinear*, type.

Such a linear word can be represented in the form of an oriented *graph* of the binary-tree type, as in the case of Paragraph 1.2.2 [eq. (1-6)]. It is seen on the graph that each letter of such a word has one antecedent, except the first letter (the "root" of the tree) which has none. As a consequence of (1-26): on the graph each T, L or R is followed by one letter, each B or C by two letters; after an S no letter follows; the number of S exceeds by one the number of B and C.

A generating word of the type commented on in Paragraph 1.2.4 [eqs. (1-12) and (1-14)] is said to be *circular*. The graph has a loop, which expresses the presence of *feedback* from a certain letter towards the first. In this case the (1-26) conditions are not fulfilled anymore. Each letter (including the first letter) has an antecedent.

We shall speak of a *global* loop if the feedback originates in the last letter of the word (as in the case of SD_{D1}) and of a *local* loop if it originates in another letter of the word (as in the case of SD_{D2}).

B) Generating word without an operating system

In developmental systems the generating word of which has no operating system all the $A_i(a_i)$ operations are performed simultaneously at each developmental step; an (n,n) matrix can be associated to the graph. We shall call it the *evolution matrix* of the system. It is most easily constructed on the basis of the elementary operations acting on each cell category: see the last two columns of Figure 1-15.

The evolution matrix of a developmental system without feedback is upper-triangular. Its elements are zeros and ones. The ones are located in lines which express respectively:

a) a stagnation (S), if the one lies in the main diagonal;

b) a transformation (T), if the one lies at one or two spaces at the right of the main diagonal;

c) a linear (L) or rotative (R) generation, if a one lies in the main diagonal and a second one lies at one or two spaces more to the right;

d) a bifurcation with (C) or without (B) change of direction, if two ones lie close to each other at the right of the main diagonal.

Hence a "map" of the evolution matrix (Figure 1-16) can be drawn, in which stagnations give rise to ones on the main diagonal, passive operators (T, B, C) to ones at the right thereof, and active operators (L, R) to pairs of ones located on and at the right of the main diagonal. (For the terms "passive" and "active", see § 2.1.1-F *in fine*.) A one located in the lower-left half of the matrix expresses the presence of feedback, as in the case of Examples DS_{D1} and DS_{D2}.

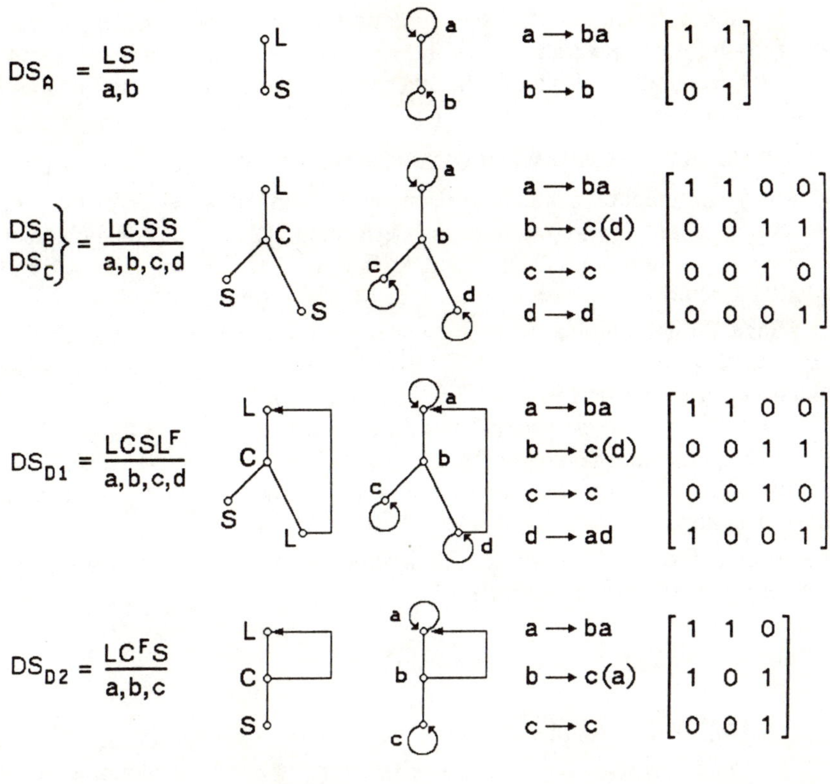

Fig. 1-15. Graph and evolution matrix of developmental systems DS_A to DS_D.

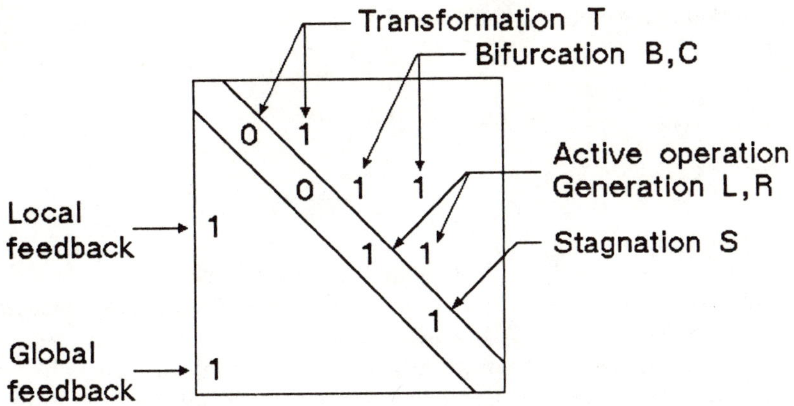

Fig. 1-16. Location of the ones in the evolution matrix of a developmental system.

It should be noted that the evolution matrix makes no distinction between the two generative operations L and R, nor between the two bifurcation operations B and C. In other words, the matrix ignores changes of direction.

C) Generating word with operating system

In developmental systems the generating word of which is of the type just described all $A_i(a_i)$ operations are performed (1) simultaneously at each developmental step and (2) in a context-free manner, i.e. independently of external circumstances and of the state of the system.

Such limitations can be overcome by introducing into the generating word an *operating system* (as in elaborate multiprogrammed processes) which controls the performance of the operations in such a manner that

1) they may be performed at different instants,

2) their performance may be affected by external circumstances (sensitivity to external context) or by the state of the system (sensitivity to internal context).

If O_i is the *local* operating system which controls the A_i operation, the generating word will be written

$$GW = \underset{i=a}{\overset{n}{\mathrm{SEQ}}}\, O_i A_i \tag{1-27}$$

System DS_E of Paragraph 1.2.5 is a very simple example.

A *global* operating system can also exist. It concerns the strategy of the execution of the A_i operations as a whole:

$$GW = O_G \left\{ \underset{i=a}{\overset{n}{\mathrm{SEQ}}}\, A_i \right\} \tag{1-28}$$

Still more generally, the operating system can have both global and local parts:

$$GW = O_G \left\{ \underset{i=a}{\overset{n}{\mathrm{SEQ}}}\, O_i A_i \right\} \tag{1-29}$$

*

We are not, by far, the only authors who write on the subject of developmental systems. Our starting point was Aristid Lindenmayer's publications (1968 to 1976) which gave rise to an important body of literature on *L systems*, sometimes at a high level of generality and abstraction (for example, Rozenberg and Salomaa, 1980). Spencer Brown's *laws of forms* have been applied to biological phenomena (Varela, 1979) and Thom's *catastrophe theory* is known to a large public (1977). We would also like to quote the work by Roger Jean (1978) devoted to the mathematical description of plant growth.

Our aim has been to develop a mathematically simple model compatible with the fundamental assumptions of cellular biology. In fact such a model leads to easily interpretable results without any mathematical difficulty and, despite its simplicity, accounts for many biological facts. It is essentially based on the concept of a *generating word*, which is the analog of the program of a computer and of the genetic code of DNA present in the cells of a living organism

The following three chapters will be devoted to developmental systems with increasing degree of complexity: synchronous developmental systems without feedback (Chap. 2), with feedback (Chap. 3) and developmental systems with an operating system (Chap. 4). The systems will be investigated from the double viewpoint of their structural properties and of their quantitative growth.

REFERENCES

J.C. GILLE, S. WĘGRZYN and P. VIDAL (1988), On some models for developmental systems. Part IX: Generalized generating word and genetic code, *International Journal of Systems Science*, vol. 19, pp. 845-855.

J.L. HALL, T.J. FLOWERS and R.M. ROBERTS (1974), *Plant cell structure and metabolism*, Longman, London, 426 p.

Roger JEAN (1978), *Phytomathématique*, Presses de l'Université du Québec, Montréal, 271 p.

A. LINDENMAYER (1968), Mathematical models for cellular interactions in development, *Journal of Theoretical Biology*, vol. 18, pp. 280-315.

A. LINDENMAYER (1971), Developmental systems without cellular interactions, their languages and grammars, *Journal of Theoretical Biology*, vol. 30, pp. 455-484.

A. LINDENMAYER (1975), Developmental systems and languages in their biological context in G.T. HERMAN and G. ROZENBERG, *Developmental Systems and Languages*, North-Holland Publishing Co., Amsterdam and American Elsevier, New York, pp. 1-40.

A. LINDENMAYER and G. ROZENBERG (eds.) (1976), *Automata. Languages. Development. At the crossroads of biology, mathematics and computer science*, North-Holland Publishing Co., Amsterdam and American Elsevier, New York, 529 p.

H.M. MARTINEZ (1976), Automaton-theoretic models of cellular development *in* Lindenmayer, A. and Rozenberg, G. (eds.) (1976), pp. 125-137.

G. ROZENBERG and A. SALOMAA (1980), *The Mathematical Theory of L Systems*, Academic Press, New York, 352 p.

R. THOM (1977), *Stabilité structurelle et Morphogénèse: essai d'une théorie générale des modèles*, 2nd ed., Union générale d'éditions, Paris, 351 p.

F.G. VARELA (1979), *Principles of biological autonomy*, North-Holland, New York, 306 p.

P. VIDAL, S. WĘGRZYN and J.C. GILLE (1978), Sur la notion de système évolutif, *Bulletin de l'Académie polonaise des Sciences (Série des sciences techniques)*, vol. 26, pp. 387-390.

P. VIDAL, S. WĘGRZYN and J.C. GILLE (1982), Systèmes évolutifs: opérations ponctuelles et applications biologiques et techniques, *Revue française d'automatique, d'informatique et de recherche opérationnelle (Automatique)*, vol. 16, pp. 147-173.

P.M.B. VITÁNYI, Digraphs associated with D0L systems (1976) *in* Lindenmayer, A. and Rozenberg, G. (eds.) (1976), p. 335-346.

S. WĘGRZYN, J.C. GILLE and P. VIDAL (1982), On some models for developmental systems. Part I: Organism structure on the basis of elementary transformations, *International Journal of Systems Science*, vol. 13, pp. 359-378.

S. WĘGRZYN, P. VIDAL and J.C. GILLE (1981), Système évolutif défini par des opérations ponctuelles sur ses composantes, *Bulletin de l'Académie polonaise des Sciences (Série des sciences techniques)*, vol. 29, pp. 317-320.

NOTE

A great part of our research on developmental systems is summarized in the following articles:

S. WĘGRZYN, J.C. GILLE and P. VIDAL (1988), Modele matematyczne procesów rozwoju [Mathematical Models of Developmental Processes], *Kosmos*, vol. 37, pp. 3-46.

S. WĘGRZYN, J.C. GILLE and P. VIDAL (1989), A model for developmental systems. Part I: Generating word without an operating system, *Automatica*, vol. 25, pp. 695-706.

J.C. GILLE, S. WĘGRZYN and P. VIDAL (1989), A model for developmental systems. Part II: Generating word with operating system, *Automatica*, vol. 25, pp. 707-714.

A first short version of the present book was published in Polish with the title *Genetyka procesów rozwoju* [The Genetics of Developmental Processes], Polytechnic of Silesia Press, Gliwice, 1988, 70 p.

Chapter 2

SYNCHRONOUS DEVELOPMENTAL SYSTEMS WITHOUT FEEDBACK

The present chapter is devoted to the simplest kind of developmental systems: systems the generating word of which has no operating system (i.e. all operations are performed simultaneously and indefinitely) and no feedback (i.e. the first cell has no antecedent). The generating word of such a system has the shape of a binary tree. Examples were the systems DS_A, DS_B and DS_C of Section 1.2.

We shall first explain, with the help of examples of systems constructed mainly with C, L and S operations, in what manner the development is controlled by the generating word. Then structural and finally quantitative properties will be investigated.

2.1 GENERATING WORD AND SYSTEM DEVELOPMENT: ANALYSIS AND SYNTHESIS

2.1.1 System with stem and branches. Role of L

Example C of Paragraph 1.2.3

$$DS_C = \frac{L\,C\,S\,S}{a,b,c,d} \tag{1-11}$$

has the shape of an indefinitely growing tree with branches of unit length. In the present paragraph emphasis will be laid on the role played by the L operation in the growth of the system.

A) REMOVING L

Removing L yields the developmental system

$$DS_F = \frac{C_{\pm\alpha}\,S\,S}{a,b,c} \tag{2-1}$$

It is easily seen that the sequence of operations

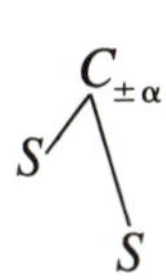

$$a \rightarrow b(c) \qquad b \rightarrow b \qquad c \rightarrow c \tag{2-2}$$

produces a development

$$DS_F(0) = a, \qquad DS_F(k) = b(c) \qquad \text{for} \qquad k > 0 \tag{2-3}$$

which does not grow indefinitely, but remains of *bounded size* (Fig. 2-1).

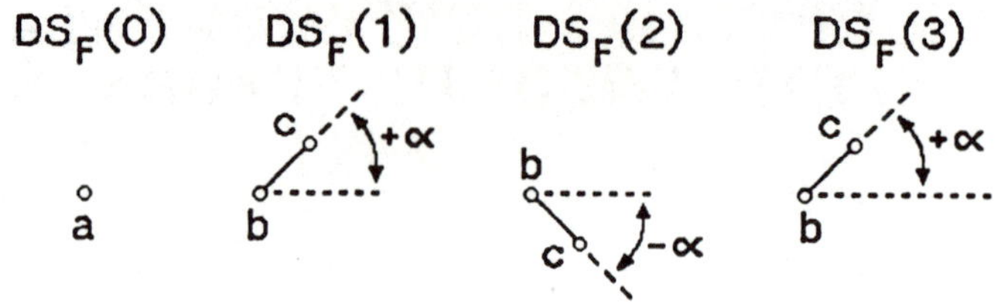

Fig. 2-1. Development of $DS_F = \dfrac{C\,S\,S}{a,b,c}$. The growth remains bounded because of the absence of any active (L,R) operation.

B) Adding L to the stem

If on the contrary a second L operation is inserted before the first S

$$DS_G = \frac{L\,C_{\pm\alpha}\,L\,S\,S}{a,b,c,d,e} \tag{2-4}$$

the sequence of operations

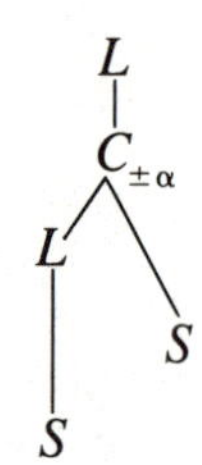

$$a \to ba \qquad b \to c(d) \qquad c \to ec \qquad d \to d \qquad e \to e \tag{2-5}$$

gives rise to the development

$DS_G(0) = a \quad DS_G(1) = ba \quad DS_G(2) = c(d)ba \quad DS_G(3) = ec(d)c(d)ba$

$DS_G(4) = eec(d)ec(d)c(d)ba \qquad DS_G(5) = eeec(d)eec(d)ec(d)c(d)ba$

$DS_G(6) = eeeec(d)eeec(d)eec(d)c(d)ba\ldots$

or, in recursive form:

$$\left.\begin{aligned} &DS_G(0) = a \qquad DS_G(1) = ba \qquad DS_G(2) = A(2)DS_G(1) \\ &\qquad\qquad\qquad \text{where} \quad A(2) = c(d) \\ &\text{and for } k > 2 \\ &\qquad\qquad DS_G(k) = eA(k)DS_G(k-1) \\ &\qquad\qquad\quad A(k) = eA(k-1) \end{aligned}\right\} \quad (2\text{-}6)$$

which has the shape of a tree with unit-length branches; but, as a consequence of the additional L, the interbranch distance increases at each step (Fig. 2-2), i.e. the *stem grows faster*.

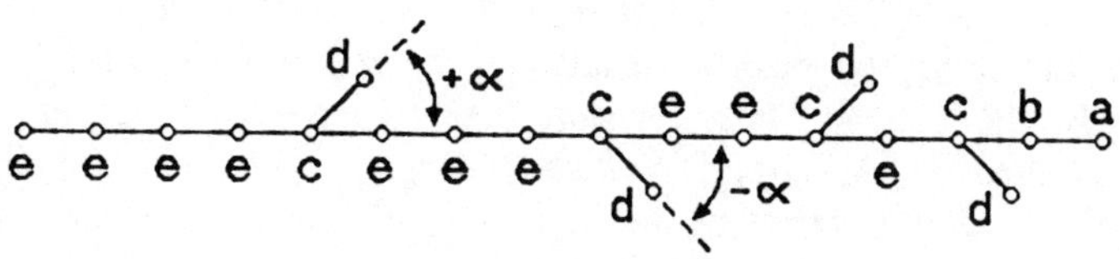

Fig. 2-2. Development of system $DS_B = \dfrac{LCLSS}{a,b,c,d,e}$.

C) Adding L to the branches

If the additional L operation is inserted before the second S

$$DS_H = \frac{LC_{\pm\alpha}SLS}{a,b,c,d,e} \quad (2\text{-}7)$$

the sequence of operations

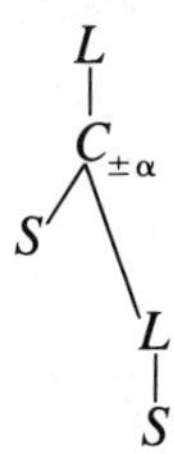

$$\begin{aligned} a &\rightarrow ba \\ b &\rightarrow c(d) \\ c &\rightarrow c \\ d &\rightarrow ed \\ e &\rightarrow e \end{aligned} \quad (2\text{-}8)$$

produces the development

$$\left.\begin{array}{c} DS_H(0) = a \qquad DS_H(1) = ba \qquad DS_H(2) = c(d)ba \\ DS_H(3) = c(ed)c(d)ba \qquad DS_H(4) = c(eed)c(ed)c(d)ba \\ DS_H(5) = c(eeed)c(eed)c(ed)c(d)ba \\ DS_H(6) = c(eeeed)c(eeed)c(eed)c(ed)c(d)ba \\ DS_H(7) = c(eeeeed)c(eeeed)c(eeed)c(eed)c(ed)c(d)ba \end{array}\right\} \quad (2\text{-}9a)$$

or, in recursive form

$$\left.\begin{array}{l} DS_H(3) = c(A(2))DS_H(2) \qquad \text{where} \qquad A(2) = ed \\ \text{and for } k > 2 \\ DS_H(k) = c(A(k-1))DS_H(k-1) \end{array}\right\} \quad (2\text{-}9b)$$

The development has the shape of a tree whose *branches grow indefinitely* (Fig. 2-3). The *meristem* (see Paragraphs 1.2.3 and 1.3.2) consists of the a, b and d cells, which undergo division; the remainder of the system (c and e cells) is the analog of the conduction and support tissues of a plant ("carrier" subsystem).

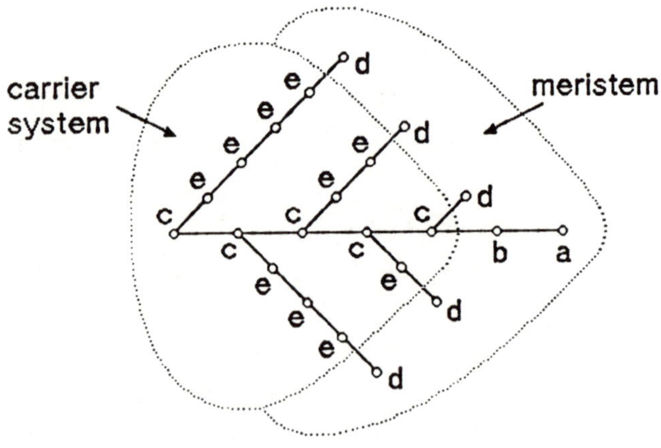

Fig. 2-3. Development of $DS_H = \dfrac{LCSLS}{a,b,c,d,e}$: indefinitely growing branches.

D) Adding L to stem and to branches

If now two L operations are added

$$DS_I = \frac{L\,C_{\pm\alpha}\,L\,S\,L\,S}{a,b,c,d,e,f} \tag{2-10}$$

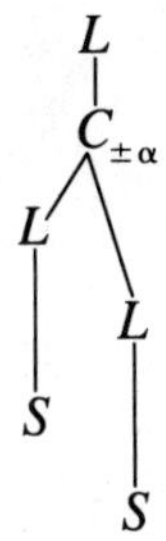

$$\begin{aligned} a &\to ba \\ b &\to c(d) \\ c &\to ec \\ d &\to fd \\ e &\to e \\ f &\to f \end{aligned} \tag{2-11}$$

it is easily seen that the two above effects are combined: the branches grow indefinitely and the interbranch distance increases. Figure 2-4 shows

$$DS_I(6) = eeeec(ffffd)eeec(fffd)eec(ffd)ed(fd)c(d)ba \tag{2-12}$$

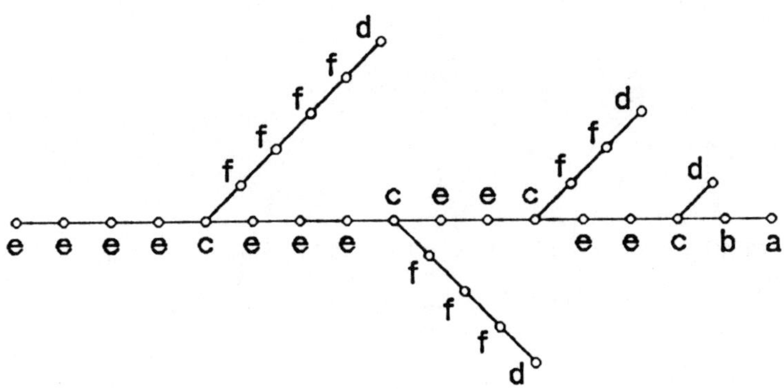

Fig. 2-4. Development of system $DS_I = \dfrac{LCLSLS}{a,b,c,d,e,f}$.

E) Linear generation L versus transformation T

If T is substituted to L in DS_C (eq. 1-11)

$$DS_J = \frac{T\,C_{\pm\alpha}\,S\,S}{a,b,c,d} \tag{2-13}$$

the sequence of operations

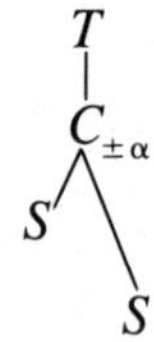

$$a \to b \qquad b \to c(d) \qquad c \to c \qquad d \to d \tag{2-14}$$

produces a development

$$\left.\begin{aligned} DS_J(0) &= a \qquad DS_J(1) = c(d) \\ DS_J(k) &= c(d) \qquad \text{for} \qquad k > 1 \end{aligned}\right\} \tag{2-15}$$

which remains indefinitely bounded.

Similarly if T is substituted to the second L of the system DS_H (indefinitely growing branches, eq. 2-8):

$$DS_K = \frac{L\,C_{\pm\alpha}\,S\,T\,S}{a,b,c,d,e} \tag{2-16}$$

the sequence of operations

$$a \to ba \qquad b \to c(d) \qquad c \to c \qquad d \to e \qquad e \to e \tag{2-17}$$

gives rise to a system the branches of which remain of unit length (Fig. 2-5):

$$DS_K(6) = c(e)c(e)c(e)c(e)c(d)ba \tag{2-18}$$

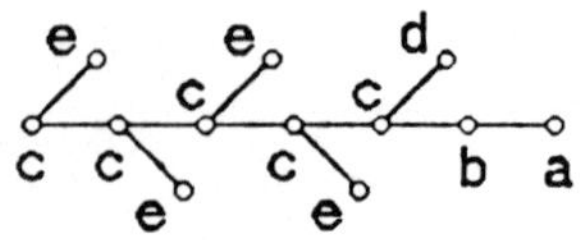

Fig. 2-5. Developmental stage $k = 6$ of $DS_K = \dfrac{LCSTS}{a,b,c,d,e}$.

Similarly the reader will check that replacing the last two L of DS_{I} (eq. 2-10) by T

$$DS_L = \frac{L\,C_{\pm\alpha}\,T\,S\,T\,S}{a,b,c,d,e,f} \tag{2-19}$$

yields a tree with unit-length branches and constant interbranch distance.

F) LINEAR GENERATION L VERSUS ROTATIVE GENERATION R

If an R operation is substituted to the second L of DS_H (eq. 2-7)

$$DS_L = \frac{L\,C_{\pm 45}\,S\,R_{\alpha}\,S}{a,b,c,d,e} \tag{2-20}$$

the $d \rightarrow ed$ operation (eq. 2-8) is replaced by $d \rightarrow e(d)$:

$$DS_L(6) = c(e(e(e(e(d)))))c(e(e(e(d))))c(e(e(d)))c(e(d))c(d)ba \tag{2-21}$$

As a consequence the branches grow spirally (Fig. 2-6) but the system has not been affected from the topological viewpoint.

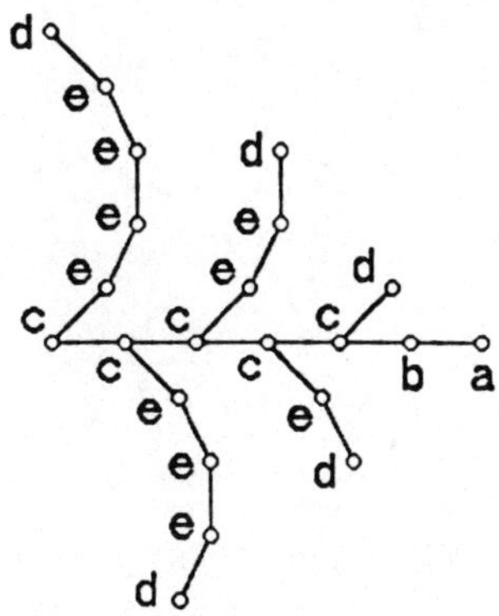

Fig. 2-6. Developmental stage $k = 6$ of $DS_L = \dfrac{LCSRS}{a,b,c,d,e}$.

It results from the above (especially from Paragraphs A, E and F) that the growth of a system or of parts thereof is due to the presence of L and R operations. For this reason they will be called *active* operations, as opposed to the *passive* operations T, B and C.

2.1.2 Branches and subbranches. Role of C

The comparison of developmental systems DS_A and DS_B (see Section 1.2) shows that a C operation is responsible for the generation of branches.

A) ADDING C TO BRANCHES

If a second C operation is added to the generating word of DS_H (eq. 2-8), the following system is obtained

$$DS_M = \frac{L\,C\,S\,L\,C\,S\,S}{a,b,c,d,e,f,g,h} \tag{2-22}$$

where both C are to be understood as $C_{\pm 45}$.

The sequence of operation becomes

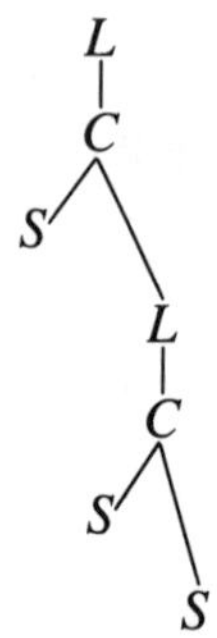

$$\begin{aligned} a &\rightarrow ba \\ b &\rightarrow c(d) \\ c &\rightarrow c \\ d &\rightarrow ed \\ e &\rightarrow f(g) \\ f &\rightarrow f \\ g &\rightarrow g \end{aligned} \tag{2-23}$$

whence the development

$$\begin{gathered} DS_M(0) = a \qquad DS_M(1) = ba \qquad DS_M(2) = c(d)ba \\ DS_M(3) = c(ed)c(d)ba \qquad DS_M(4) = c(f(g)ed)c(ed)c(d)ba \\ DS_M(5) = c(f(g)f(g)ed)c(f(g)ed)c(ed)c(d)ba \end{gathered} \tag{2-24}$$

$$DS_M(6) = c(f(g)f(g)f(g)ed)c(f(g)f(g)ed)c(f(g)ed)c(ed)c(d)ba$$

The indefinitely growing branches now bear subbranches of finite length. Figure 2-7 shows the $k = 7$ developmental stage.

On the basis of the above, more complicated development systems can be synthesized.

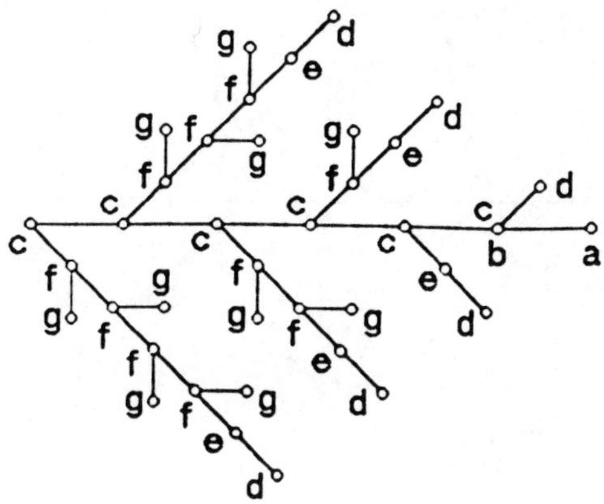

Fig. 2-7. Developmental stage $k = 7$ of $DS_M = \dfrac{L\,C\,S\,L\,C\,S\,S}{a,b,c,d,e,f,g}$: branches and subbranches.

a) Adding an L after the second C of DS_M causes the subbranches to grow indefinitely:

$$DS_N = \frac{L\,C\,S\,L\,C\,S\,L\,S}{a,b,c,d,e,f,g,h} \tag{2-25}$$

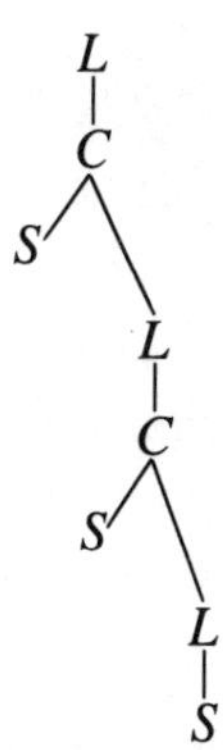

$$\begin{aligned} a &\to ba \\ b &\to c(d) \\ c &\to c \\ d &\to ed \\ e &\to f(g) \\ f &\to f \\ g &\to hg \\ h &\to h \end{aligned} \tag{2-26}$$

The $k = 6$ developmental stage is

$$DS_N(6) = c(f(hhg)f(hg)ed)c(f(hg)\,f(g)ed)c(f(g)ed)c(ed)c(d)ba \tag{2-27}$$

Stage $k = 7$ is shown in Figure 2-8.

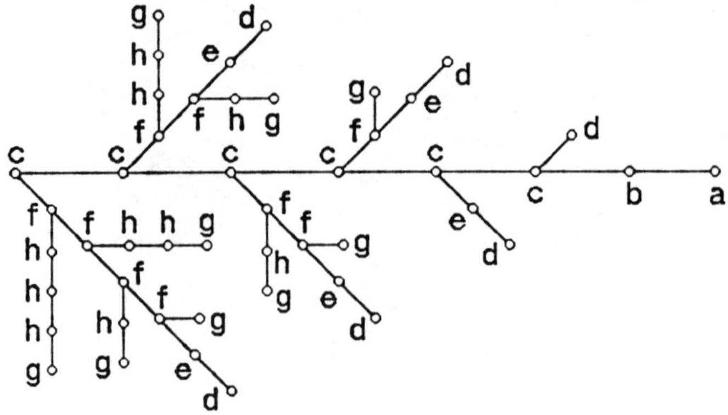

Fig. 2-8. Developmental stage $k = 7$ of $DS_N = \dfrac{L\,C\,S\,L\,C\,S\,L\,S}{a,b,c,d,e,f,g,h}$.

b) The reader will verify that adding a third C

$$DS_O = \frac{L\,C\,S\,L\,C\,S\,L\,C\,S\,S}{a,b,c,d,e,f,g,h,i} \tag{2-28}$$

produces sub-subbranches of finite length and adding a third L

$$DS_P = \frac{L\,C\,S\,L\,C\,S\,L\,C\,S\,L\,S}{a,b,c,d,e,f,g,h,i,j} \tag{2-29}$$

causes the sub-subbranches also to grow indefinitely.

B) Removing C

Conversely a developmental system whose generating word has no C (and no R) operation develops only in one direction. Such was the case for DS_A (eq. 1-2) of Paragraph 1.2.1. Such is also, e.g., the case for

$$DS_P = \frac{L\,B\,S\,S}{a,b,c,d} \tag{2-30}$$

where the C of (1-11) has been replaced by a B.

The reader will easily check that

$$DS_P(6) = cdcdcdcdcdba \tag{2-31}$$

which confirms that the C operation is responsible for the system developing in two dimensions.

2.1.3 Symmetrical patterns

A) Principle

The branch configuration of Figures 2-2 to 2-8 differs from the symmetry which characterizes many living organisms, at least as a first approximation.

It can be easily seen that symmetry can be accounted for by introducing dichotomy into the elementary operations of the generating word (Gille, Węgrzyn and Vidal, 1988).

Consider the two symmetric developmental systems

$$X = \frac{L\,C_{\alpha}\,S\,L\,S}{i,j,k,l,m} \qquad \overline{X} = \frac{L\,C_{-\alpha}\,S\,L\,S}{i,j,k,l,m}$$

adapted from DS_H (eq. 2-7) and differing from each other by the fact that the C bifurcation of the X system implies a change of direction α but that of the $\overline{X}$ system implies the change of direction $-\alpha$. We shall write, using parentheses for the former and square brackets for the latter(¹):

$$\begin{cases} C_{\alpha} & j \rightarrow k(l) \\ C_{-\alpha} & j \rightarrow k[l] \end{cases}$$

The development of the X and $\overline{X}$ systems is thus defined by

X		$\overline{X}$	
$i \rightarrow ji$	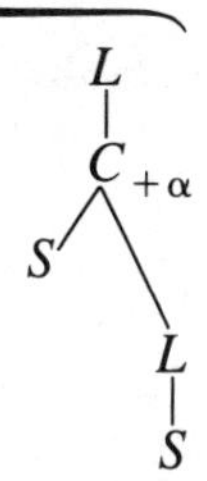	$i \rightarrow ji$	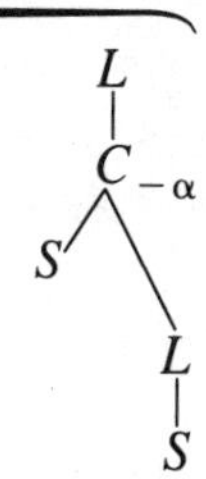
$j \rightarrow k(l)$		$j \rightarrow k[l]$	
$k \rightarrow k$		$k \rightarrow k$	
$l \rightarrow ml$		$l \rightarrow ml$	
$m \rightarrow m$		$m \rightarrow m$	

and their $k = 5$ developmental stage is respectively

$$X(5) = k(mmml)k(mml)k(ml)k(l)ji$$

$$\overline{X}(5) = k[mmml]k[mml]k[ml]k[l]ji$$

(¹) It should be mentioned that as early as 1975 G.T. Herman and G. Rozenberg (p. 50), after having written "we have not bothered to distinguish between the sides on which the branches may lie", immediately add: "this could have been done by the use of different types of brackets."

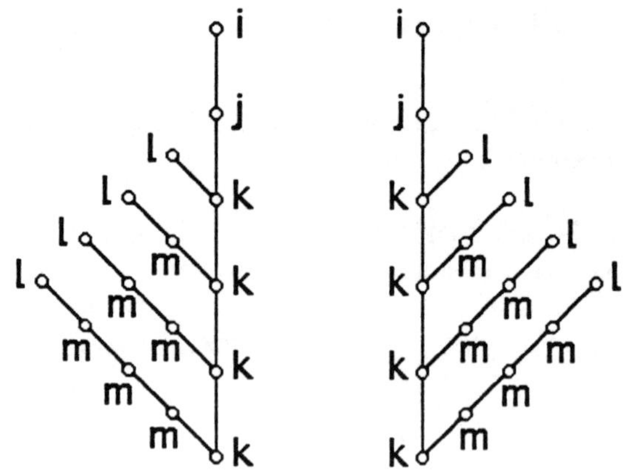

Fig. 2-9. Symmetrical development of two systems (k = 5).

The symmetry between the development of the X and $\overline{X}$ systems appears on the latter expressions and on Figure 2-9. If by some means X and $\overline{X}$ are properly combined as subsystems of a composite system, the latter will evince symmetry. One manner of implementing such a combination will be shown in the next paragraph.

B) Example

Now consider the developmental system shown in Figure 2-10 where the X and $\overline{X}$ subsystems are those of the foregoing section and the Y subsystem is defined as (Fig. 2-11)

$$Y = \frac{L\,S}{r,s} \qquad \begin{matrix} L \\ | \\ S \end{matrix} \qquad \begin{matrix} r \rightarrow sr \\ s \rightarrow s \end{matrix}$$

The whole system can be described as

$$DS = \frac{B\,B\,B\,C\,C\,T\,T\,S\,C\,T\,S}{a,b,c,e_1,e_2,X_k,\overline{X}_k,f,d,Y_k,g}$$

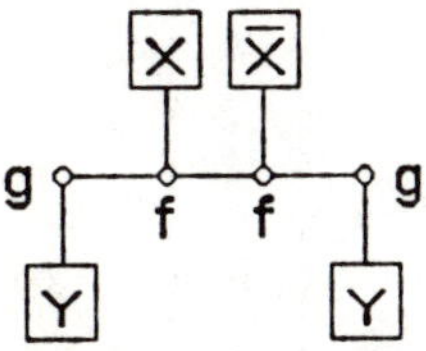

Fig. 2-10. Complex developmental system made up of symmetric subsystems.

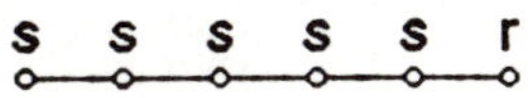

Fig. 2-11. $k = 8$ developmental stage of subsystem $Y = \dfrac{LS}{r,s}$.

the operations being (²)

$a \rightarrow bc$	B	$X_k \rightarrow X_{k+1}$	T
$b \rightarrow de_1$	B	$\overline{X}_k \rightarrow \overline{X}_{k+1}$	T
$c \rightarrow e_2 d$	B	$f \rightarrow f$	S
$e_1 \rightarrow f \backslash X_k /$	C	$d \rightarrow g / Y_k \backslash$	C
$e_2 \rightarrow f \backslash \overline{X}_k /$	C	$Y_k \rightarrow Y_{k+1}$	T
		$g \rightarrow g$	S

where the horizontal direction from left to right is considered as the main developmental direction and the perpendicular directions are denoted by the symbols \/ (below to above) and /\ (above to below) respectively.

Symmetry is obvious in Figure 2-12, which shows the $k = 8$ developmental stage

$$DS(8) = g / ssssssr \backslash f \backslash k(mmml)k(mml)k(ml)k(l)ji / f \backslash k$$
$$[mmml]k[mml]k[ml]k[k]ji / g / ssssssr \backslash$$

(²) Here T should be understood as a generalized T operation in the sense that it is applied not to a single cell (eq. 1-21), but to all the cells of a subsystem X_k or $\overline{X}_k$: each cell goes over from its state at the k-th developmental stage to its state at the $(k + 1)$th stage.

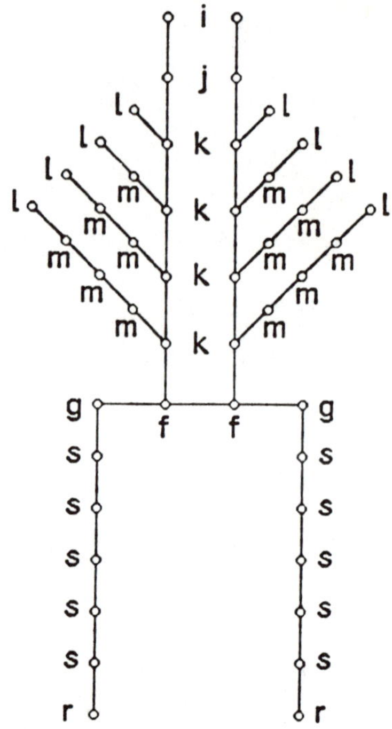

Fig. 2-12. $k = 8$ developmental stage of the symmetric system of Figure 2-10.

2.2 STRUCTURAL PROPERTIES

Some structural properties of developmental systems without feedback will now be investigated. They are a consequence of the binary-tree shape of the graph of the generating word and of the resulting hierarchy that exists inside the system (Gille, Węgrzyn and Vidal, 1984; Węgrzyn, Gille and Vidal, 1984; Węgrzyn, Vidal and Gille, 1985).

2.2.1 Hierarchical decomposition

A) General

The generating word of a developmental system is made up of letters, each of which symbolizes an elementary operation. Now it often happens that a certain group of letters inside the generating word has enough autonomy and enough importance to deserve being considered as an autonomous entity, as a *subword*, which is a part of the whole word, in particular is actuated by the operations (letters) preceding it. The relation of such a subword to an isolated letter S, T, L, R, B, C is the same as the relation of a *macroinstruction* to a simple instruction in a computer language. Such a subword can be considered as the generating word of a developmental subsystem: the subsystem that it generates is a *pattern* inside the whole developmental system.

The presence of subsystems having a common pattern within biological systems ("morphogenetic repetition") has been noted since long by biologists (e.g. Maresquelle and Sell, 1965). A. Lindenmayer (1975, p. 34-35) and D. Frijters and A. Lindenmayer (1976) discussed patterns inside a developmental system and studied recursive relations between the developmental stages thereof, but they did not elaborate the concept of a subword inside a generating word.

The purpose of the present paragraph is to show in what manner a generating graph can be broken down into subgraphs, then the subgraphs into sub-subgraphs, etc. down to the stagnation symbols — and to suggest an analogy with an organism which consists of organs: organs contain tissues; tissues are made up of cells (Gille, Węgrzyn and Vidal, 1984; Vidal, Węgrzyn and Gille, 1984; Węgrzyn, Gille and Vidal, 1988, p. 89-91).

B) Hierarchical decomposition of a system

A developmental system DS is described by a generating word GW that is a sequence of operations A_i acting on the cells a_i

$$GW = A_a A_b \ldots A_m = \operatorname*{SEQ}_{i=a}^{m} A_i \qquad \text{(1-18)}$$

The first letter can be isolated by writing

$$GW = A_a \operatorname*{SEQ}_{i=b}^{m} A_i \qquad \text{(2-32)}$$

If A_a is an L, an R or a T, i.e., an operation without bifurcation, the SEQ at the right hand of (5) can be treated as one subword X

$$GW = A_a X \qquad \text{(2-33a)}$$

If A_a is a B or a C, i.e., a bifurcation, the word sequence consists of two subwords X and Y:

$$GW = A_a \genfrac{}{}{0pt}{}{\diagup X}{\diagdown Y} \qquad \text{(2-33b)}$$

The procedure described by (2-33a) or (2-33b) can then be applied to the first letter of X (or of X and Y), etc. in a recursive manner until the remaining sequence X or Y is nothing but an S (see Figure 2-13).

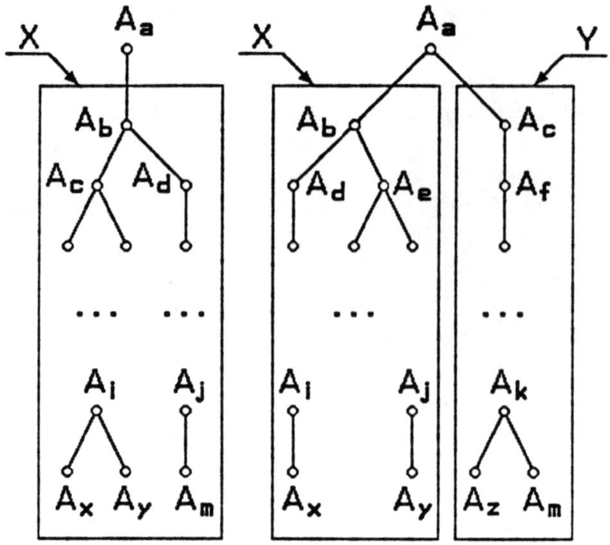

Fig. 2-13. Hierarchical breakdown of a generating word without feedback according to whether the first operation A_a is T, L, R (left: A_aX) or is B, C (right: A_aXY). Note: the indexes of the terminal A (stagnations) on the last line cannot be specified *a priori*, except the last one which is m.

More precisely, we have the following (see Figure 2-14).

(*a*) If the first operation A_a is an L (linear generation), the development of the system starting from one a cell as its initial condition is described as

$$a \rightarrow DS_x(k)a$$

$$DS_x(k) \rightarrow DS_x(k+1)$$

where the motive DS_x is the developmental subsystem whose generating word is X.

Hence

$$k = 0,\ DS(0) = a$$

$$\mathrm{k} > 0,\ DS(k) = \underset{\mathrm{i}=1}{\overset{\mathrm{k}-1}{\mathrm{SEQ}}}\, DS(k-1)\, a$$

In other words the system generated by LX consists of a *chain of subsystems* whose *generating word is X*.

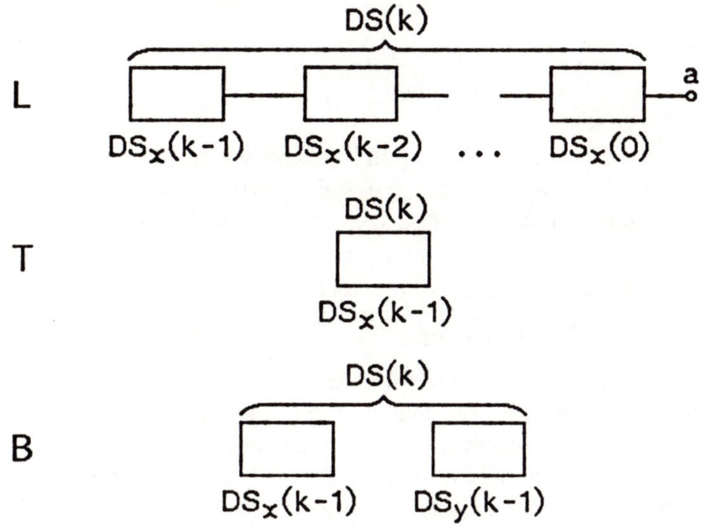

Fig. 2-14. Result of the action of operation L, T on subsystem X and of operation B on subsystems X and Y.

If A_a is an R (rotative generation) the same holds, except that the chain has a spiral form.

(*b*) If A_a is a T (differentiation), the development is described by

$$a \rightarrow DS_x(k)$$

$$DS_x(k) \rightarrow DS_x(k+1)$$

whence

$$k = 0, \quad DS(0) = a$$

$$k > 0, \quad DS(k) = DS_x(k-1)$$

i.e., the system generated by TX has the same development as X but *shifted* by one developmental stage.

(*c*) If A_a is a B (bifurcation) the operations are

$$a \rightarrow DS_x(k)DS_y(k)$$

$$DS_x(k) \rightarrow DS_x(k+1)$$

$$DS_y(k) \rightarrow DS_y(\mathrm{k}+1)$$

whence

$$k = 0, \quad DS(0) = a$$

$$k > 0, \quad DS(k) = DS_x(k-1)DS_y(k-1)$$

In other words the developmental system generated by *BXY* consists of two motives DS_x and DS_y the *generating word of which are X and Y* respectively and which are shifted with respect to each other by one stage.

If A_a is a *C* (bifurcation with change of direction) the same holds, except that the shift is oblique in direction.

2.2.2 "Tissues" and "organs"

When obtaining the successive developmental stages of a system numerically it often appears that a certain group of letters within the generating word has enough autonomy and enough importance to merit consideration as an autonomous entity. This is because a subword, which is a part of the whole word, is actuated in particular by the operations (letters) preceding it. As was stated in the foregoing paragraph, such a subword can be considered as being the generating word of a developmental subsystem, of a *pattern* inside the whole developmental system.

In a similar manner sub-subwords can sometimes be isolated within a subword and they generate sub-subsystems inside subsystems.

Such a *hierarchy* may be compared with the hierarchy that exists in complex organisms: the whole organism consists of systems of organs; organs contain tissues; tissues are made up of cells.

Example 1. First consider the extremely simple case

$$GW = LX \qquad X = S$$

If the initial conditions are $DS(0) = a$, $DS_x(0) = b$ the development occurs as follows:

$$k = 0,\ DS(0) = a$$

$$k = 1,\ DS(1) = ba$$

$$k > 1,\ DS(k) = \underbrace{bb\ldots b}_{k}a$$

Figure 2-15 shows the $k = 7$ stage: it can be interpreted as a tissue consisting of *b* cells.

If *X* is slightly less simple, e.g.

$$GW = LX \qquad X = CSS$$

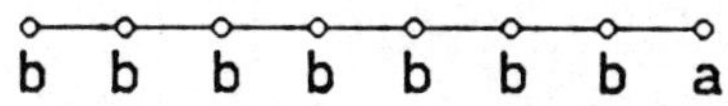

Fig. 2-15. System *LS* at stage $k = 7$: rectilinear "tissue" of b cells.

the development which starts from the initial conditions $DS(0) = a$, $DS_x(0) = b$, $DS_x(1) = c(d)$ is

$$k = 0, \; DS(0) = a$$

$$k = 1, \; DS(1) = DS_x(0) = ba$$

$$k > 1, \; DS(k) = DS_x(k-1)\dots DS_x(1)DS_x(0)a = \underbrace{c(d)c(d)\dots c(d)}_{k}ba$$

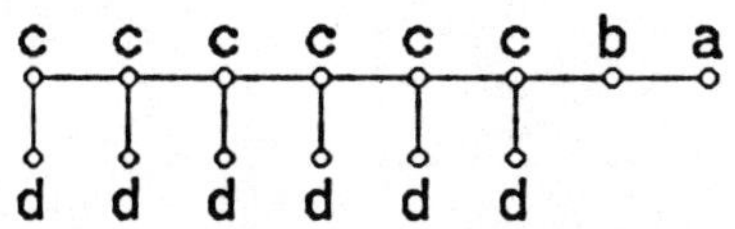

Fig. 2-16. System *LCSS* at stage $k = 7$: rectilinear "tissue" of c and d cells.

Figure 2-16 shows the $k = 7$ stage: a "tissue" consisting of cellular couples $c(d)$. If L is replaced by R

$$GW = RX \qquad X = CSS$$

one obtains a similar "tissue", but with a spiral form. The $k = 11$ stage is shown in Figure 2-17.

Example 2. Consider the more complex developmental system DS_Q, the generating word of which is

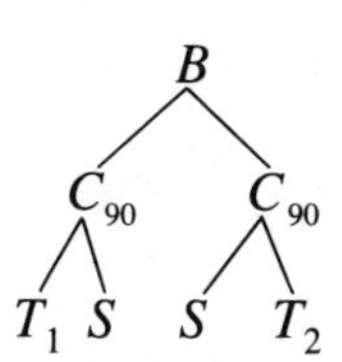

(2-34)

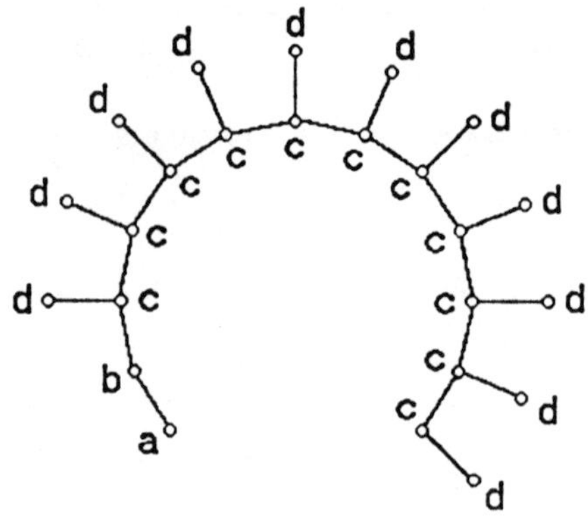

Fig. 2-17. System *RCSS* at stage $k = 14$: spiral "tissue" of c and d cells.

where the subsystems T_1 and T_2 are respectively

$$T_1 = \frac{L\,C_{90}\,S\,S}{d,e,f,g} \qquad\qquad T_2 = \frac{L\,C_{-90}\,S\,S}{h,i,f,g}$$

The sequence of operations is:

$$\begin{array}{llll} a \to bc & d \to ed & h \to ih & f \to f \\ b \to m(d) & \underbrace{e \to f(g)}_{T_1} & \underbrace{i \to f[g]}_{T_2} & g \to g \\ c \to m(h) & & & \\ m \to m & & & \end{array}$$

The $k = 6$ developmental stage (Fig. 2-18)

$$DS_Q = m(f(g)f(g)f(g)ed)m(f[g]f[g]f[g]ih) \tag{2-35}$$

can be looked upon as an "organ" ([3]) consisting of two "tissues".

2.2.3 Cloning. Grafting

It is possible to combine two generating words by placing the "root" of one graph at an external node of the other (Węgrzyn, Vidal and Gille, 1981; Węgrzyn, Gille and Vidal, 1986). This makes it possible to "construct" complex developmental systems. Such a recombination is the

---

([3]) What is meant here by "organ" is a composite tissue. In Physiology an organ is primarily defined by its function (*causa finalis*) and secondly by its cellular structure (*causa formalis*), but our model only takes the latter into account.

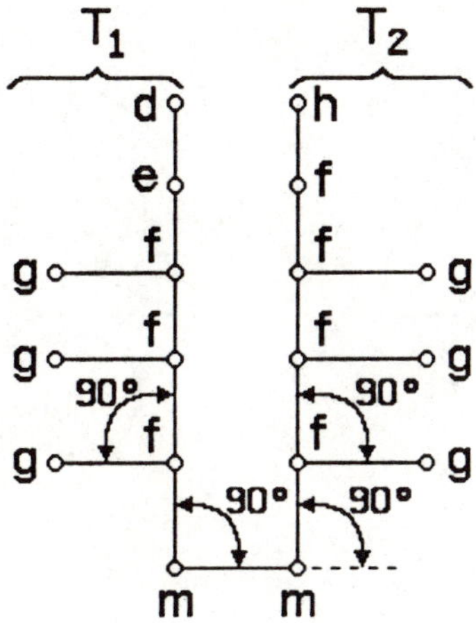

Fig. 2-18. System (2-34) at stage $k = 6$: rectilinear "tissues" T_1 and T_2 and m, m cells.

image of the procedure of *cloning*, which consists of inserting a fragment of DNA into the genome of a host cell, thus resulting in a combination of genes.

Consider the developmental system (Example C of Paragraph 1.2.3)

$$DS_C = \frac{GW_C}{Z_C} = \frac{LCSS}{a,b,c,d} \tag{1-11}$$

and suppose the $P = LCSS$ fragment of another generating word extended over the $Z = \{d,e,f,g\}$ cell set is introduced into GW_C at the last S. The result of this cloning procedure is the more complex system

$$DS_C' = \frac{LCSP}{a,b,c,Z} = \frac{LCSLCSS}{a,b,c,d,e,f,g}$$

The $k = 7$ developmental stages are respectively (see Figure 2-19)

$$DS_C(7) = c(d)c(d)c(d)c(d)c(d)c(d)ba$$

$$DS_C'(7) = c(f(g)f(g)f(g)f(g)ed)c(f(g)f(g)f(g)ed)c(f(g)f(g)ed)$$
$$c(f(g)ed)c(ed)c(d)ba$$

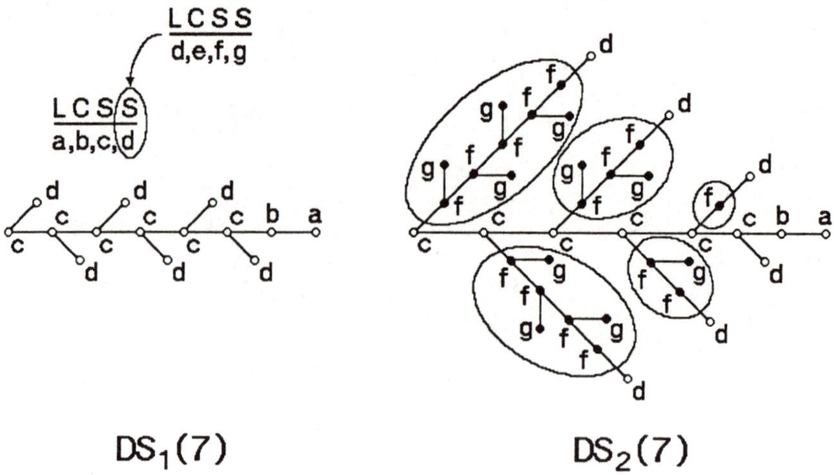

Fig. 2-19. Developmental system before (left) and after (right) cloning.

Note the difference between cloning and *grafting* (Gille, Węgrzyn and Vidal, 1989). The latter procedure consists of extracting a fragment of an organism and inserting it into another organism.

If for example the third d cell of the $k = 4$ stage of the above DS_C system

$$DS_C(4) = c(d)c(d)c(d)ba$$

is replaced by the initial cell e of the developmental system

$$\frac{LCSS}{e,f,g,h}$$

a new system is obtained

$$DS_C'' = \frac{LCSSLCSS}{a,b,c,d,e,f,g,h}$$

with the initial condition

$$DS_C''(0) = c(d)c(d)c(e)ba$$

The $k = 4$ developmental stage

$$DS_C''(4) = c(d)c(d)c(g(h)g(h)g(h)fe)c(d)c(d)c(d)c(d)ba$$

corresponds to the $k = 8$ developmental stage of the original system DS_C (prior to grafting): see Figure 2-20.

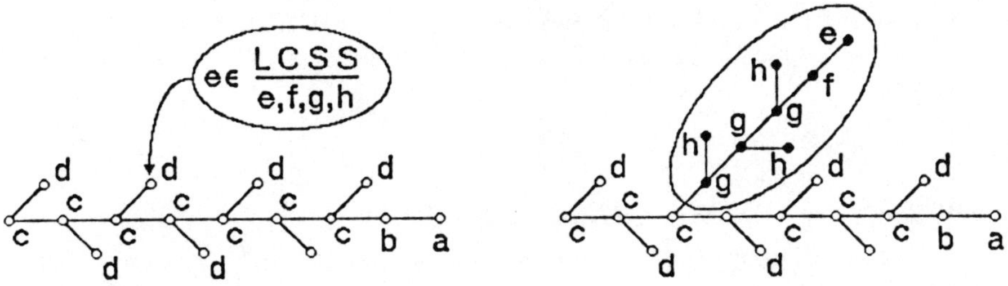

Fig. 2-20. Developmental system before (left) and after (right) grafting.

2.2.4 Influence of initial conditions on development: regeneration

Consider the five-letter generating word of Figure 2-21. If the initial state is one *a* cell the five cell categories will be present after the $k = 5$ developmental stage and will undergo the associated operations. But if the initial condition is one *d* or *e* cell, then no *a*, *b* or *c* cell will ever appear. If the initial condition is one *b* cell, no *a* cell will ever be found.

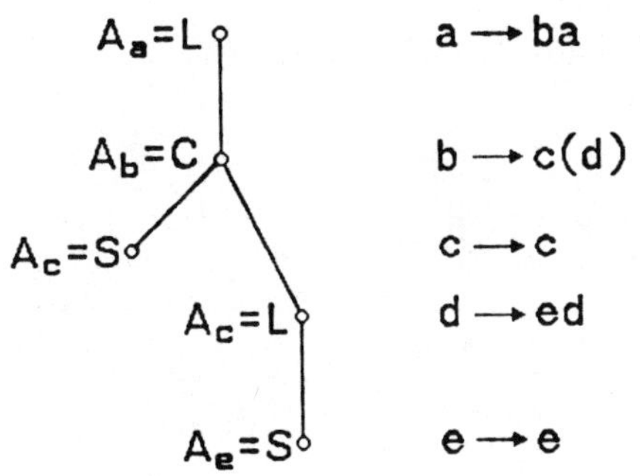

Fig. 2-21. System *LCSLS*.

More generally if the generating word is rectilinear the system will develop its full structure only if the initial condition consists of *a* cell(s) associated to the first operation of the generating word, i.e. to the root of the graph. If the initial condition consists of i cell(s) ($i = b,c,\ldots,m$) only the pattern generated by the subword the initial letter of which is A_i

$$\underset{j=i}{\overset{m}{\mathrm{SEQ}}}\, A_j \tag{2-37}$$

will develop (Gille, Węgrzyn and Vidal, 1986).

For the system shown in Figure 2-22 the correspondence is given in Table 2.1

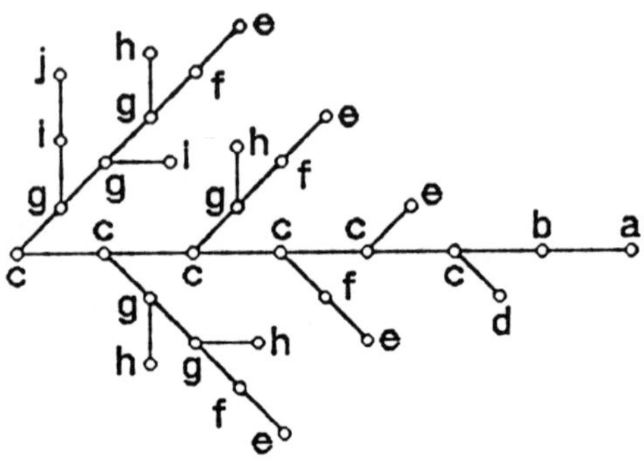

Fig. 2-22. System *LCSTLCSTLS* (see eq. 2-38) at stage $k = 9$.

Table 2.1. Subsystems parts of the system shown in Figure 2-17 which develop depending on the initial state.

Initial condition	$Z(k),\ k > m$
a	$A_aA_bA_cA_dA_e$
b	$A_bA_cA_dA_e$
c	A_c
d	A_dA_e
e	A_e

This fact has an important consequence concerning the possibility of regeneration of a developmental system with a rectilinear generating word.

Suppose that all the cells of such a system have been destroyed except the stagnant cells (i.e. the cells associated with stagnation operations S: they constitute a sort of permanent structure in the sense that they do not develop any more and are independent of the generating word) and except one i cell associated with the A_i operation. As a consequence of the above *only the subsystem generated by the subword (2-37) having its root at* A_i *will be regenerated*. Depending on the location of the said operation in the generating word a more or less important portion of the system will regenerate, but not the entire system (except of course if the surviving cell is an a cell associated with the first operation A_a of the generating word).

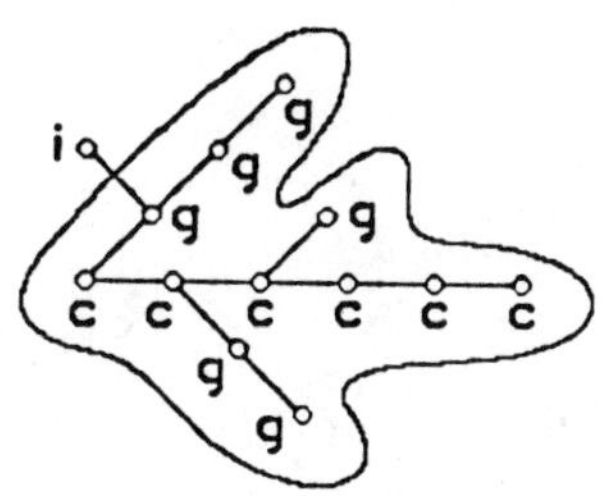

Fig. 2-23. Partial destruction of system (2-38): only the stagnant cells and one i cell survive.

For example the structure of the $k = 9$ developmental stage of the system

$$
\begin{aligned}
a &\rightarrow ba \\
b &\rightarrow c(d) \\
c &\rightarrow c \\
d &\rightarrow e \\
e &\rightarrow fe \\
f &\rightarrow g(h) \\
g &\rightarrow g \\
h &\rightarrow i \\
i &\rightarrow ji \\
j &\rightarrow j
\end{aligned}
\qquad (2\text{-}38)
$$

is shown in Figure 2-22.

Now suppose that all the cells have been destroyed with the exception of the stagnant cells and of one i cell (see Figure 2-23), belonging to the meristem. Only the subsystem

$$\begin{array}{cc} L & \qquad i \to ji \\ | & \\ S & \qquad j \to j \end{array}$$

will develop, i.e., only a "tissue" consisting of j cells will regenerate (see Figure 2-24).

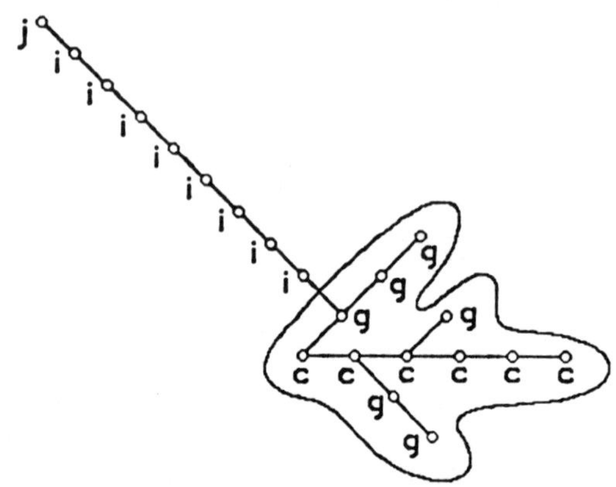

Fig. 2-24. Regeneration of system (2-38) after the partial destruction shown in Figure 2-23.

2.2.5 Mutation

Suppose that a change occurs in the generating word of a developmental system in the sense that one operation A_i is replaced by a different one A'_i. In order that the conditions (1-26) be still satisfied it is necessary that:

(a) a T, L or R operation be changed to another operation which can only be followed by one letter (i.e., a T, L or R, not a B or C);

(b) a B or C operation be changed to another operation which can only be followed by two letters (i.e., a C or B, not a T, L or R);

(c) S operations remain S.

If the generating word is rectilinear it is clear as a consequence of the principle of hierarchy (§ 2.2.2) that such a change will *only affect the subsystem generated by the subword (2-37) having its root at the modified letter* A_i (Węgrzyn, Vidal and Gille, 1985 and 1986).

This will be shown on one example. Consider the generating word

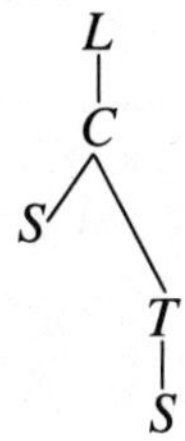

$a \to ba$

$b \to c(d)$

$c \to c$

$d \to e$

$e \to e$

and suppose the T operation is replaced by an L, so that the modified system (the *mutant*) is

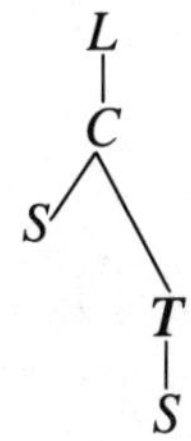

$a \to ba$

$b \to c(d)$

$c \to c$

$\boldsymbol{d \to ed}$ [modified operation]

$e \to e$

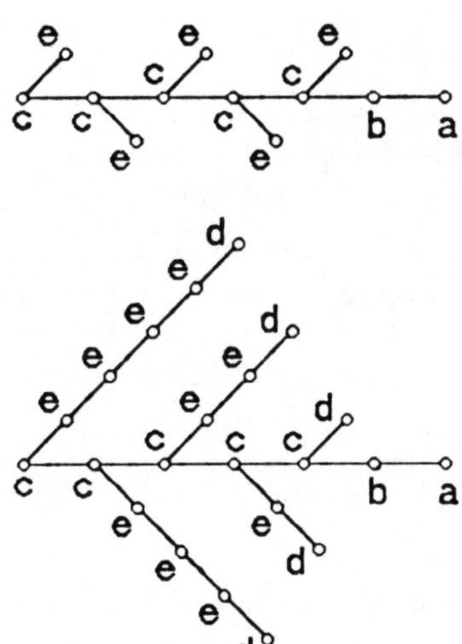

Fig. 2-25. Top: system *LCSTS* at stage $k = 6$. Bottom: the same after the substitution (mutation) $T \to L$.

If the initial condition is one a cell, the $k = 7$ developmental stage is for the initial system

$$DS(6) = c(e)c(e)c(e)c(e)c(d)ba$$

and for the modified system

$$DS_m(6) = c(eeeed)c(eeed)c(eed)c(ed)c(d)ba$$

It is apparent in Figure 2-25 that the mutation does not alter the structure of the whole system (a tree with equally spaced branches) but only affects the branches: instead of having finite length they now grow indefinitely.

2.3 QUANTITATIVE GROWTH

2.3.1 General

The volume of a developmental system varies in time (growth of the "organism") as well as its cellular composition (cells of certain categories proliferate, others disappear). The growth has given rise to a great number of studies (Szilard, 1973; Paz and Salomaa, 1973; Berstel and Nielsen, 1976; Herman and Vitányi, 1976; Rozenberg and Salomaa, 1976 and 1980; Salomaa, 1976 and 1981; Soittola, 1976; Vitányi, Karhumaïki, Ehrenfeucht and Rozenberg, 1981; see also: Herman and Rozenberg, 1975, pp. 269-283; Salomaa and Soittola, 1978, pp. 95-117), often carried out at a high level of abstraction and generality. We shall show that the growth and the cellular composition of the developmental systems considered here can be analyzed by very simple methods.

Given a developmental system the generating word of which consists of n operations, the numbers of the cells of each category at the k-th developmental stage will be respectively noted $a(k)$, $b(k)$, ..., $n(k)$, the *cellular composition* being the row matrix

$$\mathbf{Y}(k) = [a(k) \quad b(k) \quad \dots \quad n(k)] \tag{2-39}$$

The total number of cells is the *size* or *volume* of the system

$$V(k) = a(k) + b(k) + \dots + n(k) \tag{2-40}$$

Obviously $V(k)$ is the product of $\mathbf{Y}(k)$ by the column matrix consisting of n ones:

$$V(k) = \mathbf{Y}(\mathrm{k}) \; [1 \quad 1 \quad \dots \quad 1]^T \tag{2-41}$$

2.3.2 Numerical approach

The numerical values of $a(k)$, …, $n(k)$ and of $V(k)$ can be obtained by computing $DS(k)$ at each k on the basis of the elementary operations (§ 1.3.1). From this it is sometimes possible (in the case of very simple systems) to obtain the analytical expression of $\mathbf{Y}(k)$ and $V(k)$.

Such was the case for DS_A (eq. 1-5) and for DS_B or DS_C (eq. 1-10).

For DS_K it is immediately inferred from the successive $DS(k)$ (see eq. 2-18) that

$$V(k) = 2k \tag{2-42}$$

In the case of systems DS_G, DS_H and DS_L (which have the same quantitative behavior) it is apparent from (2-6) that

$$\left.\begin{aligned} &a(k) = 1 \qquad b(k) = u(k-1) \qquad c(k) = d(k) = (k-1)u(k-2) \\ &e(k) = [1+2+3+\ldots+(k-2)]u(k-3) = \tfrac{1}{2}(k^2-3k+2)u(k-3) \end{aligned}\right\} \tag{2-43}$$

where $u(x)$ is the unit-step function

$$u(x) = \begin{cases} 0 & x<0 \\ 1 & x \geqslant 0 \end{cases} \tag{2-44}$$

Hence for $k \geqslant 3$

$$V(k) = 1 + 1 + 2(k-1) + \frac{k^2-3k+2}{2} = \frac{k^2+k+2}{2} \tag{2-45}$$

For system DS_I consideration of the successive $DS(k)$ (see eq. 2-12) shows that

$$\left.\begin{aligned} &a(k) = b(k) = 1 \qquad c(k) = d(k) = (k-2)u(k-2) \\ &e(k) = f(k) = [1+2+\ldots+(k-2)]u(k-3) = \tfrac{1}{2}(k^2-3k+2)u(k-3) \end{aligned}\right\} \tag{2-46}$$

whence

$$V(k) = k^2 + k + 2 \tag{2-47}$$

For system DS_N considering the successive $DS(k)$ (see eq. 2-24) yields

$$\left.\begin{aligned} a(k) = b(k) = 1 \qquad c(k) = d(k) = (k-1)u(k-2) \\ e(k) = (k-2)u(k-3) \\ f(k) = g(k) = [1+2+\ldots+(k-3)]u(k-4) = \tfrac{1}{2}(k^2-5k+6)u(k-4) \end{aligned}\right\} \tag{2-48a}$$

and (less easily)

$$g(k) = \tfrac{1}{6}(k^3 - 9k^2 + 26k - 24)u(k-5) \tag{2-48b}$$

whence for $k \geqslant 5$

$$V(k) = \tfrac{1}{6}(k^3 - 3k^2 + 14k) \tag{2-49}$$

But in most cases the mathematical form of $\mathbf{Y}(k)$ and $V(k)$ can be found only be resorting to a matrix or to a transform approach.

2.3.3 Matrix approach

A) GENERAL

Consideration of the *evolution matrix* **M** of the developmental system is especially adequate for a quantitative evaluation of $\mathbf{Y}(k)$ and $V(k)$. Recall that **M** is the matrix associated to the graph that expresses the generating word in explicit form. Each row of the matrix corresponds to one level (one operation) of the graph.

The following examples are self-explanatory.

$$\begin{aligned} a &\to ba \\ b &\to c(d) \\ c &\to c \\ d &\to e(d) \\ e &\to e \end{aligned} \qquad \mathbf{M} = \begin{bmatrix} 1 & 1 & 0 & 0 & 0 \\ 0 & 0 & 1 & 1 & 0 \\ 0 & 0 & 1 & 0 & 0 \\ 0 & 0 & 0 & 1 & 1 \\ 0 & 0 & 0 & 0 & 1 \end{bmatrix}$$

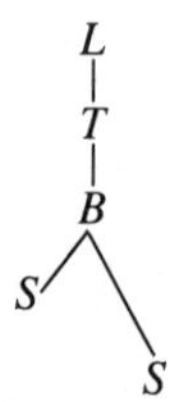

$$\begin{aligned} a &\to ba \\ b &\to c \\ c &\to de \\ d &\to d \\ e &\to e \end{aligned} \qquad \mathbf{M} = \begin{bmatrix} 1 & 1 & 0 & 0 & 0 \\ 0 & 0 & 1 & 0 & 0 \\ 0 & 0 & 0 & 1 & 1 \\ 0 & 0 & 0 & 1 & 0 \\ 0 & 0 & 0 & 0 & 1 \end{bmatrix}$$

The manner in which the location of the ones is related to the operations which compose the generating word was explained in Paragraph 1.3.2-B.

The system transition from state $DS(k-1)$ to state $DS(k)$ is a consequence of the operations of the generating word, which are applied to $DS(k-1)$. Quantitatively the cellular composition is affected according to

$$\mathbf{Y}(k) = \mathbf{Y}(k-1)\,\mathbf{M} \tag{2-50}$$

The reader will easily verify this relation with the help of the examples DS_A, DS_B and DS_C (§ 1.2, Fig. 1-10) or of DS_F (eq. 2-3), DS_G (2-6), DS_H (2-9), DS_M (2-24).

As a consequence the cellular composition at the k-th developmental stage is given in terms of the initial cellular composition $\mathbf{Y}(0)$ by

$$\boxed{\mathbf{Y}(k) = \mathbf{Y}(0)\,\mathbf{M}^k} \tag{2-51}$$

and the system size is (see eq. 2-41)

$$V(k) = \mathbf{Y}(0)\,\mathbf{M}^k\,[1 \quad 1 \quad \dots \quad 1]^T \tag{2-52}$$

The initial state is usually assumed to be one cell of the first category, i.e.

$$\mathbf{Y}(0) = [1 \quad 0 \quad 0 \quad \dots \quad 0]^T \tag{2-53}$$

Then

$$\mathbf{Y}(k) = \text{first row of } \mathbf{M}^k \tag{2-54}$$

$$V(k) = \text{sum of elements of first row of } \mathbf{M}^k \tag{2-55}$$

Analyzing the evolution of the cellular composition and of the size of the system thus essentially consists of evaluating $\mathbf{M}^k$ as a function of k. According to eqs. (2-50), (2-51) and (2-55) the evolution matrix $\mathbf{M}$ of the system could also be termed its *growth matrix*.

B) Direct computation of $\mathbf{M}^k$

In some very simple cases $\mathbf{M}^k$ is easily obtained directly.

For DS_A (§ 1.2.1)

$$\mathbf{M} = \begin{bmatrix} 1 & 1 \\ 0 & 1 \end{bmatrix} \qquad \mathbf{M}^k = \begin{bmatrix} 1 & k \\ 0 & 1 \end{bmatrix}$$

whence by (2-54) and (2-55)

(1-4b, 1-5) $\quad \mathbf{Y}(k) = [1 \quad k] \qquad V(k) = k + 1$

For DS_B or DS_C (§ 1.2.2, 1.2.3)

$$\mathbf{M} = \begin{bmatrix} 1 & 1 & 0 & 0 \\ 0 & 0 & 1 & 1 \\ 0 & 0 & 1 & 0 \\ 0 & 0 & 0 & 1 \end{bmatrix} \qquad \mathbf{M}^k = \begin{bmatrix} 1 & 1 & k-1 & k-1 \\ 0 & 0 & 1 & 1 \\ 0 & 0 & 1 & 0 \\ 0 & 0 & 0 & 1 \end{bmatrix}$$

whence by (2-54) and (2-55)

(1-9b, 1-10) $\quad \mathbf{Y}(k) = [1 \quad 1 \quad k-1 \quad k-1] \qquad V(k) = 2k$

C) Use of the Cayley-Hamilton theorem

In most cases obtaining $\mathbf{M}^k$ is not so easy (Winiarczyk, 1981). Using the Cayley-Hamilton theorem for evaluating growth functions, as was proposed by A. Salomaa (1973), is advantageous in the present case of a developmental system with a rectilinear generating word (Gille, Vidal, Węgrzyn and Ouellet, 1982; Gille, Vidal and Węgrzyn, 1983).

As a consequence of the Cayley-Hamilton theorem (for example, Gantmacher, 1959, pp. 83, 113-116) $\mathbf{M}^k$ can be, however great k may be, obtained as a linear combination of the identity matrix $\mathbf{I}_n$ and of the first $(n - 1)$ powers of $\mathbf{M}$:

$$\mathbf{M}^k = \mu_0\mathbf{I}_n + \mu_1\mathbf{M} + \mu_2\mathbf{M}^2 + \ldots + \mu_{n-1}\mathbf{M}^{n-1}$$

where the μ are scalar quantities that depend on k.

Premultiplying by $\mathbf{Y}(0)$ one obtains

$$\mathbf{Y}(0)\mathbf{M}^k = \mu_0\mathbf{Y}(0) + \mu_1\mathbf{Y}(0)\mathbf{M} + \mu_2\mathbf{Y}(0)\mathbf{M}^2 + \ldots + \mu_{n-1}\mathbf{Y}(0)\mathbf{M}^{n-1}$$

whence, taking (2-51) into account:

$$\mathbf{Y}(k) = \mu_0\mathbf{Y}(0) + \mu_1\mathbf{Y}(1) + \mu_2\mathbf{Y}(2) + \ldots + \mu_{n-1}\mathbf{Y}(n-1) \qquad (2\text{-}56)$$

and from (2-52)

$$V(k) = \mu_0V(0) + \mu_1V(1) + \mu_2V(2) + \ldots + \mu_{n-1}V(n-1)$$

In other words, once $\mathbf{Y}(1)$, $\mathbf{Y}(2)$, ..., $\mathbf{Y}(n-1)$ have been numerically computed (step by step, as was shown on several examples in sections 1.2, 2.1 and 2.2) $\mathbf{Y}(k)$ is obtained for any subsequent k as a linear combination of them... provided the μ are known.

From the viewpoint of developmental processes this fact can be looked upon as a consequence of the idea that the genetic code of an organism — which has completely "come to light" after the n-th developmental step — determines all its future context-free development.

In the present case of a developmental system without feedback this method has practical value because the evolution matrix $\mathbf{M}$ is triangular and therefore its characteristic values are the figures found on the main diagonal, i.e. 0 and 1 (see Figure 1-16). The order of multiplicity of 0 is the number p of the B, C and T operations; the order of multiplicity of 1 is the number q of the L, R and S operations. The caracteristic polynomial of $\mathbf{M}$ is thus

$$P(\lambda) = \lambda^p(\lambda - 1)^q$$

where p and q are the numbers of the B, C, T and L, R, S operations respectively (denoted by small letters):

$$p = b + c + t \qquad q = l + r + s \qquad p + q = n$$

$$P(\lambda) = \lambda^{b+c+t}(\lambda - 1)^{l+r+s}$$

According to a classical theory of linear algebra (e.g. Gille and Clique, 1988, p. 10-23) the first p coefficients μ

$$\mu_0, \mu_1, \mu_2, \ldots, \mu_{p-1}$$

are obtained by substituting $\lambda = 0$ into the equation

$$\lambda^k = \mu_0 + \mu_1\lambda + \mu_2\lambda^2 + \ldots + \mu_{n-1}\lambda^{n-1}$$

and into the equations obtained by differentiating it $(p - 1)$ times with respect to λ. Adding the supplementary condition $k \geqslant p$ (in order that all the left-hand sides be zero), this yields:

$$\mu_0 = 0 \quad \mu_1 = 0 \quad \mu_2 = 0 \quad \ldots, \quad \mu_{p-1} = 0$$

The remaining q coefficients μ

$$\mu_p, \mu_{p+1}, \mu_{p+2}, \ldots, \mu_{p+q-1} = \mu_{n-1}$$

are obtained in a similar manner by substituting $\lambda = 1$ into the same equation

$$\lambda^k = \mu_p\lambda^p + \mu_{p+1}\lambda^{p+1} + \ldots + \mu_{n-1}\lambda^{n-1}$$

(recall that $\mu_0 = \mu_1 = \ldots = \mu_{p-1} = 0$) and into the equations obtained by differentiating it $(q - 1)$ times with respect to λ. This yields for any positive k, after some algebraic manipulation:

$$\mu_i = \frac{(k-p)(k-p-1)\ldots(k-i+1)(k-i-1)\ldots(k-n+1)}{(i-p)(i-p-1)\ldots(1)(-1)\ldots(i-n+1)}$$

$$\mu_i = \prod_{\substack{j=p \\ j\neq i}}^{n-1} \frac{k-j}{i-j} \qquad i = p, p+1, \ldots, n-1$$

i.e. $\mu_i(k)$ is the Lagrange interpolation polynomial of degree $(q - 1)$ which is equal to one for $k = i$ and to zero for $k = p, p+1, \ldots, i-1, i-1, \ldots, n-1$.

Hence the procedure for obtaining the cellular composition of the system at its k-th stage of development $(k \geqslant p)$ is the following:

a) evaluate $\mathbf{Y}(1), \mathbf{Y}(2), \ldots, \mathbf{Y}(n - 1)$ numerically,

b) then obtain $\mathbf{Y}(k)$ from (2-56), in which the first p coefficients μ are zero and the last q ones are given by the above formula.

Example. For the system

$$\text{(2-7)} \qquad DS_H = \frac{L\,C\,S\,L\,S}{a,b,c,d,e}$$

one has

$$b = 0 \quad c = 1 \quad t = 0 \quad l = 2 \quad r = 0 \quad s = 2$$

whence

$$p = 1 \qquad q = 4 \qquad P(\lambda) = \lambda(\lambda - 1)^4$$

The first five $\mathbf{Y}(k)$ are (see equations 2-9a)

a	$\mathbf{Y}(0) = [1 \quad 0 \quad 0 \quad 0 \quad 0]$
ba	$\mathbf{Y}(1) = [1 \quad 1 \quad 0 \quad 0 \quad 0]$
c(d)ba	$\mathbf{Y}(2) = [1 \quad 1 \quad 1 \quad 1 \quad 0]$
c(ed)c(d)ba	$\mathbf{Y}(3) = [1 \quad 1 \quad 2 \quad 2 \quad 1]$
c(eed)c(ed)c(d)ba	$\mathbf{Y}(4) = [1 \quad 1 \quad 3 \quad 3 \quad 3]$

Hence

$$\mu_0 = 0$$

$$\mu_1 = \frac{(k-2)(k-3)(k-4)}{(1-2)(1-3)(1-4)} = \frac{-k^3+9k^2-26k+24}{6}$$

$$\mu_2 = \frac{(k-1)(k-3)(k-4)}{(2-1)(2-3)(2-4)} = \frac{k^3-8k^2+19k-12}{2}$$

$$\mu_3 = \frac{(k-1)(k-2)(k-4)}{(3-1)(3-2)(3-4)} = \frac{-k^3+7k^2-14k+8}{2}$$

$$\mu_4 = \frac{(k-1)(k-2)(k-3)}{(4-1)(4-2)(4-3)} = \frac{k^3-6k^2+11k-6}{6}$$

whence for $k \geqslant 1$ by (2-56), where $n = 5$:

$$\mathbf{Y}(k) = \mu_1\mathbf{Y}(1) + \mu_2\mathbf{Y}(2) + \mu_3\mathbf{Y}(3) + \mu_4\mathbf{Y}(4)$$

(2-43) $$\mathbf{Y}(k) = [1 \quad 1 \quad k-1 \quad k-1 \quad \frac{k^2-3k+2}{2}]$$

(2-45) $$V(k) = \frac{k^2+k+2}{2}$$

D) Discussion

The μ_i are polynomial functions of k. Such are also therefore $a(k)$, $b(k)$, ..., $n(k)$ and $V(k)$. The growth of the system is said to be of the *polynomial* type.

Since the μ are polynomials of degree $q-1$, it might seem at first sight that the growth occurs at a power of k equal to

$$q - 1 = l + r + s - 1$$

However this does not always hold: the highest power of k may be smaller as a consequence of algebraic simplifications. (Such is the case for the example computed in Paragraph C: degree 3 for the μ_i, degree 2 for $e(k)$, and for $V(k)$.)

The reason is the following. The above theory can be repeated with the minimal polynomial $P_{\min}(\lambda)$ of the matrix $\mathbf{M}$, which is also an annihilating polynomial. Now it may happen that the power of $(\lambda - 1)$ in the minimal polynomial is smaller than in the caracteristic polynomial:

$$P_{\min}(\lambda) = \lambda^{p'}(\lambda-1)^{q'} \qquad q' < q$$

The theory then leads to a polynomial growth with a power

$$q' - 1 < l + r + s - 1$$

Note that only the first $p' + q'$ $(< n)$ developmental stages need to be computed.

The minimal polynomial of **M** cannot be written immediately. But it is possible readily to write

$$P_{\text{int}}(\lambda) = \lambda^{t+b+c}(\lambda - 1)^{l+r} = \lambda^{p}(\lambda - 1)^{q+1-s}$$

which can be shown to be an annihilating polynomial of **M**. If there is more than one stagnation S in the generating word (i.e., if there is at least one bifurcation operation B or C) its degree is lower than the degree of the characteristic polynomial. The said polynomial is then so to speak "intermediate" between the characteristic polynomial and the minimal polynomial (in some cases it is identical to the latter). Using it for obtaining $\mathbf{Y}(k)$ and $V(k)$ leads to μ functions the degree of which is

$$q - s = l + r$$

and demands that only $n + 1 - s$ ($< n$) developmental stages be numerically computed.

Example. For the developmental system DS_H considered above (Paragraph C)

$$p = 1 \qquad q + 1 - s = 3 \qquad P_{\text{int}}(\lambda) = \lambda(\lambda - 1)^3$$

(In this particular case $P_{\text{int}}(\lambda)$ is precisely the minimal polynomial.) Hence

$$\mu_0 = 0 \qquad \mu_1 = \frac{(k-2)(k-3)}{(1-2)(1-3)} = \frac{k^2 - 5k + 6}{2}$$

$$\mu_2 = \frac{(k-1)(k-3)}{(2-1)(2-3)} = -k^2 + 4k - 3 \qquad \mu_3 = \frac{(k-1)(k-2)}{(3-1)(3-2)} = \frac{k^2 - 3k + 2}{2}$$

(Note the degree 2.)

$\mathbf{Y}(k)$ and $V(k)$ for $k \geqslant 1$ are obtained on the basis of $\mathbf{Y}(0)$, $\mathbf{Y}(1)$, $\mathbf{Y}(2)$ and $\mathbf{Y}(3)$ only:

$$\mathbf{Y}(k) = \mu_0\mathbf{Y}(0) + \mu_1\mathbf{Y}(1) + \mu_2\mathbf{Y}(2) + \mu_3\mathbf{Y}(3)$$
$$V(k) = \mu_0 V(0) + \mu_1 V(1) + \mu_2 V(2) + \mu_3 V(3)$$

Of course, the same expressions are finally found as in Paragraph C *in fine*.

E) Results

The reader is referred to our two articles quoted in Paragraph C for the mathematical proofs and for a detailed discussion. Hereafter are given some general results.

1) The functions $a(k)$, $b(k)$, ..., $n(k)$, i.e. the elements of $\mathbf{Y}(k)$ and also their sum $V(k)$, are polynomial functions of k: the growth is of *polynomial* type.

2) In the *absence of any L and R* operation all the $i(k)$ ($i = a, \ldots, n$) remain finite, i.e. the growth is *bounded*.

3) The *highest power of k* which appears in the $i(k)$ and in $V(k)$ is *at most* the total *number of L and R operations* present in the generating word.

The reader should verify these results with the help of the examples computed in foregoing sections.

The size of DS_F (eq. 2-3) remains bounded because the generating word contains no active operation.

The growth occurs as the first power of k for systems DS_A (see eq. 1-4, 1-5), DS_B and DS_C (see eq. 1-9, 1-10) and also for systems DS_K (eq. 2-17) and DS_P (eq. 2-30) because only one L is present in the generating word.

The growth occurs as k^2 for systems DS_G and DS_H (see eqs. 2-6 and 2-9) whose generating words have two L. Such is also the case for DS_M (eq. 2-22).

For the system DS_N, whose generating word has three L, it was found (eq. 2-49) that the growth occurs as k^3. But for DS_I it occurs only as k^2 (eq. 2-47) although the generating word also has three L. (This difference of behavior will be clear for the reader after he has read the next paragraph.)

2.3.4 Transform approach

The easiest method for evaluating and interpreting the evolution of the cellular composition of a developmental system consists in obtaining the sequence

$$\{i(k)\} \qquad i(1), i(2), \ldots, i(k), \ldots \qquad i = a, \ldots, n$$

through its z-transform

$$i(z) = i(0) + \frac{i(1)}{z} + \frac{i(2)}{z^2} + \ldots + \frac{i(k)}{z^k} + \ldots \qquad i = a, \ldots, n$$

or equivalently through its discrete Carson transform (Vidal, Węgrzyn and Gille, 1983; Winiarczyk, 1983; Węgrzyn, Gille and Vidal, 1984; Gille, Węgrzyn and Vidal, 1985).

The $\{a(k)\}$ sequence (number of the cells of the first category) is immediately obtained by observing that

1) if the first operation A_a is T, B or C, then

$$\{a_0(k)\} = 1, 0, 0, \ldots, 0, \ldots \qquad a_0(z) = 1 \tag{2-57a}$$

2) if the first operation A_a is L or R, then

$$\{a_0(k)\} = 1, 1, 1, \ldots, 1, \ldots \qquad a_0(z) = \frac{z}{z-1} \tag{2-57b}$$

Note: the subscript 0 specifies that the system has no feedback. [It will be seen later (§ 3.4.3) that $a(k)$ is modified when the generating word is circular.]

Now consider two cell categories i and j, the former immediately preceding the latter on the graph which represents the generating word. (i may be any $a, b, \ldots, m$; j may be $b, c, \ldots, n$, but not a.)

a) If the j cell is generated by a T, B or C operation A_j, then

$$j(k+1) = i(k)$$

b) If it is generated by an L, R or S operation, then

$$j(k+1) = i(k) + j(k)$$

Hence, z-transforming and noting that $j(0) = 0$ (no j cell was present at $k = 0$, since $j \neq a$):

$$\begin{cases} j(z) = \dfrac{1}{z}\, i(z) & \text{if} \quad A_j = T,B,C \qquad \text{(2-58a)} \\[2ex] j(z) = \dfrac{1}{z-1}\, i(z) & \text{if} \quad A_j = L,R,S \qquad \text{(2-58b)} \end{cases}$$

In other words, the z-transform of $\{j(k)\}$ is obtained by multiplying the z-transform of $\{i(k)\}$ by a z-transfer function $D_j(z)$

$$D_j(z) = \begin{cases} \dfrac{1}{z} & A_j = T,B,C \\[2ex] \dfrac{1}{z-1} & A_j = L,R,S \end{cases} \qquad \text{(2-59)}$$

This concept of z-transfer function $D_j(z)$ applies to any cell category except to the first cell ("root") a, which has no antecedent. However, the following notation will be introduced, for consistency's sake:

$$D_a(z) = \begin{cases} \dfrac{1}{z} & A_a = T, B, C \qquad \text{(2-60a)} \\[2ex] \dfrac{1}{z-1} & A_a = L, R \qquad \text{(2-60b)} \end{cases}$$

(The usefulness of $D_a(z)$ will become clear in the next chapter: see Paragraph 3.4.3-A.)

It results from (2-58) and (2-59) that

$$\boxed{j(z) = a(z) \prod_{\text{path}:a}^{j} D_i(z)} \qquad \text{(2-61)}$$

where the product term denotes the product of the $D_i(z)$ pertaining to the operations met *on the path leading from a to j* on the graph of the generating word, the *operation $D_a(z)$ being excluded.*

Equation (2-61) yields the explicit expression for $j(z)$ ($j = b,c,\ldots,n$) since $a_0(z)$ is known (eq. 2-57). The first operation of the generating word can be intuitively considered as a "source" which injects $a(z)$ at the input, the system consisting of the successive z-transfer functions $D_b(z)$, ..., $D_n(z)$ disposed as shown in the graph.

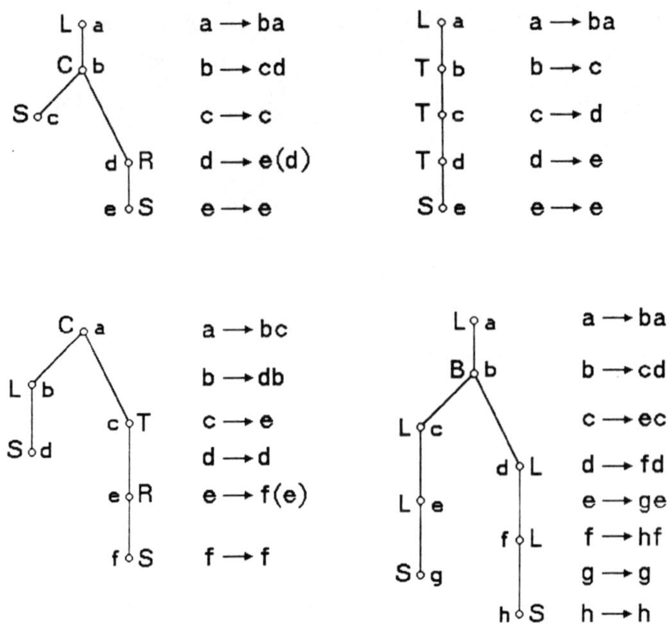

Fig. 2-26. Graph and elementary operations for the four rectilinear generating words *LCSRS*, *LTTTS*, *CLSTRS* and *LBLLSLLS*.

It should be emphasized that the product term is *not* $\prod_{i=a}^{j} D_i(z)$ in the usual sense, since the latter product would include $D_a(z)$ and *all* operations from $i = a$ to $i = j$.

Let us illustrate this by the examples shown in Figure 2-26. For the first of the four systems shown, for example

$$\prod_{\text{path:}a}^{d} D_i(z) = D_b(z)D_d(z) = \frac{1}{z}\frac{1}{z-1} \qquad [C,R] \qquad \text{(2-62)}$$

for the third

$$\prod_{\text{path:}a}^{e} D_i(z) = D_c(z)D_e(z) = \frac{1}{z}\frac{1}{z-1} \qquad [T,R] \qquad \text{(2-63)}$$

for the fourth

$$\prod_{\text{path:}a}^{h} D_i(z) = D_b(z)D_d(z)D_f(z)D_h(z) = \frac{1}{z}\frac{1}{(z-1)^3} \qquad [B,L,L,S] \qquad \text{(2-64)}$$

$$\prod_{\text{path}:a}^{e} D_i(z) = D_b(z)D_c(z)D_e(z) = \frac{1}{z}\frac{1}{(z-1)^2} \qquad [B,L,L] \tag{2-65}$$

In conclusion the sequences $\{j(k)\}$ $(j = b,\ldots,n)$ can be explicitely obtained by inverse-transforming (2-61). Here follow some general results.

If α_j and β_j are the numbers of T,B,C and of L,R operations respectively located on the graph between A_a and A_j *including* A_a, one obtains from (2-61) and (2-57)

$$J(z) = \frac{1 \text{ or } z}{z^{\alpha_j}(z-1)^{\beta_j}} \tag{2-66}$$

As a consequence of the properties of the z-transform, $j(k)$ is a *polynomial* function of k, the *degree* of which is $(\beta_j - 1)$. This fact justifies the denomination "active operation" (§ 2.1.1-F *in fine*) for L and R.

The *size* $V(k)$ of the system is also a *polynomial* function of k, the *degree* of which is

$$\sup_j \beta_j - 1 \tag{2-67}$$

i.e. is the greatest number of L, R and S operations to be found in cascade on the graph minus one — or equivalently (since the last operation of any "branch" of the graph is always an S) the *greatest number of active operations* (L,R) to be found *in cascade on the graph*.

The reader should check these results with the help of the examples computed or commented on in Paragraphs 2.3.2 and 2.3.3.

a) The absence of any active operation ($\beta_j = 0 \;\forall\; j$) causes the size of DS_F to remain bounded.

b) The growth occurs as the first power of k in the case of the systems DS_A, DS_B, DS_C, DS_K and DS_P, the generating words of which contain only one k.

c) The size of DS_G, DS_H, DS_L and DS_M grows as k^2 because their generating words have two L (or one L and one R) in cascade on the graph.

d) The generating words of systems DS_{I} and DS_N have three L. In the latter the three L but in the former only two of them are located in cascade on the graph. Therefore DS_N grows as k^3 (eq. 2-49) but DS_I grows only as k^2 (eq. 2-47), as was stated in Paragraph 2.3.3 *in fine*.

*

To summarize: the main properties of synchronous developmental systems without feedback are the following.

1) Each cell has one antecedent, except the first one, which has none.

2) The generating word has the shape of a binary tree. The evolution matrix is upper-triangular.

3) The system can be hierarchically broken down into subsystems, the subsystems into sub-subsystems, etc., in the same manner as the generating word can be broken down into subwords, sub-subwords, etc.

4) After partial destruction only a subsystem is regenerated.

5) The system is not very sensitive to mutation: changing one letter of the generating word does not alter the overall structure.

6) The growth function is of the polynomial type, with a power equal to the greatest number of active (L and R) operations found in cascade on the graph.

REFERENCES

J. BERSTEL and M. NIELSEN (1976), The growth range equivalence problem for D0L systems is decidable *in* L. LINDENMAYER and G. ROZENBERG (eds.), *Automata. Languages. Development. At the crossroads of biology, mathematics and computer science*, North-Holland Publishing Co., Amsterdam and American Elsevier, New York, pp. 161-178.

D. FRIJTERS and A. LINDENMAYER (1976), Developmental descriptions of branching patterns with paracloidal relationships *in* A. LINDENMAYER and G. ROZENBERG, *Automata. Languages. Development*, North-Holland Publishing Co., Amsterdam and American Elsevier, New York, pp. 57-73.

F.R. GANTMACHER (1959), *The Theory of Matrices*, vol. 1, Chelsea Publ. Co., New York, 374 p.

J.C. GILLE and M. CLIQUE (1988), *Calcul matriciel et Introduction à l'analyse fonctionnelle*, vol. 2, 4th ed., Lidec, Montréal, 116 p.

J.C. GILLE, P. VIDAL and S. WĘGRZYN (1987), On some models for developmental systems. Part VIII: systems with and without feedback, structural properties, *International Journal of Systems Science*, vol. 18, pp. 1195-1216.

J.C. GILLE, P. VIDAL and S. WĘGRZYN (1983), Calcul de la composition cellulaire et fonction de croissance d'un système évolutif, *Revue française d'automatique, d'informatique et de recherche opérationnelle (Automatique)*, vol. 17, pp. 87-98.

J.C. GILLE, P. VIDAL, S. WĘGRZYN and Y. OUELLET (1982), On some models for developmental systems. Part II: evolution of the cellular composition, *International Journal of Systems Science*, vol. 13, pp. 379-397.

J.C. GILLE, S. WĘGRZYN and P. VIDAL (1984), Motifs et structures dans les systèmes évolutifs, *Podstawy sterowania*, vol. 14, pp. 287-300.

J.C. GILLE, S. WĘGRZYN and P. VIDAL (1985), Grammaire des mots créateurs et propriétés des systèmes évolutifs correspondants étudiés par la transformation de Carson discrète, *Zentralblatt für Mathematik*, vol. 538, p. 350.

J.C. GILLE, S. WĘGRZYN and P. VIDAL (1986), Le principe de la régénération dans la théorie des systèmes évolutifs. Extrapolations biologiques, *Bulletin of the Polish Academy of Sciences (Technical Sciences)*, vol. 34, pp. 91-101.

J.C. GILLE, S. WĘGRZYN and P. VIDAL (1988), Dichotomie des opérations élémentaires et symétrie des systèmes évolutifs, *Bulletin of the Polish Academy of Sciences (Technical Sciences)*, vol. 36, pp. 261-264.

J.C. GILLE, S. WĘGRZYN and P. VIDAL (1989), Clonage, greffe et transplantation dans les systèmes évolutifs, *Bulletin of the Polish Academy of Sciences (Technical Sciences)*, vol. 37 [to appear].

G.T. HERMAN and G. ROZENBERG with a contribution by A. LINDENMAYER (1975), *Developmental Systems and Languages*, North-Holland Publishing Co., Amsterdam and American Elsevier, New York, 363 p.

G.T. HERMAN and P.M.B. VITÁNYI (1976), Growth function associated with biological development, *American Monthly*, vol. 83, pp. 1-15.

H.J. MARASQUELLE and Y. SELL (1965), Les problèmes physiologiques de la floraison descendante, *Bulletin de la Société française de Physiologie végétale*, vol. 11, pp. 94-98.

H.M. MARTINEZ (1976), Automaton-theoretic models of cellular development *in* A. LINDENMAYER and G. ROZENBERG (eds.), *Automata. Languages. Development*, North-Holland Publishing Co., Amsterdam and American Elsevier, New York, pp. 125-137.

A. PAZ and A. SALOMAA (1973), Integral sequential word functions and growth equivalence of Lindemayer systems, *Information and Control*, vol. 23, pp. 313-343.

G. ROZENBERG and A. SALOMAA (1976), The mathematical theory of L systems *in* J. TOU (ed.), *Advances in Information Systems Science*, Plenum Press, New York, vol. 6, pp. 161-205.

G. ROZENBERG and A. SALOMAA (1980), *The Mathematical Theory of L Systems*, Academic Press, New York, 352 p.

A. SALOMAA (1973), On exponential growth in Lindenmayer systems, *Indagationes matematicae*, vol. 35, pp. 23-30.

A. SALOMAA (1976), Growth functions of Lindermayer systems: some new approaches *in* A. LINDENMAYER and G. ROZENBERG (eds.), *Automata. Languages. Development*, North-Holland Publishing Co., Amsterdam and American Elsevier, New York, pp. 271-282.

A. SALOMAA (1981), *Jewels of Formal Language Theory*, Computer Science Press, Rockville, 144 p.

A. SALOMAA and M. SOITTOLA (1978), *Automata-theoretic aspects of formal power series*, Springer, New York, 171 p.

M. SOITTOLA (1976), Remarks on growth sequences, *Revue française d'automatique, d'informatique et de recherche opérationnelle* (*Automatique théorique*), vol. 10, pp. 23-34.

A.L. SZILARD (1971), *Growth functions of Lindenmayer systems*, Technical Report 4, Dept. of Computer Sciences, University of Western Ontario, London (Ontario), 44 p.

P. VIDAL, J.C. GILLE and S. WĘGRZYN (1986), Structure hiérarchique et principe de décomposition des systèmes évolutifs, *Bulletin of the Polish Academy of Sciences (Technical Sciences)*, vol. 34, pp. 111-119.

P. VIDAL, S. WĘGRZYN and J.C. GILLE (1983), On some models for developmental systems. Part III: another approach for evaluating the cellular composition, *International Journal of Systems Sciences*, vol. 14, pp. 753-763.

P. VIDAL, S. WĘGRZYN and J.C. GILLE (1986), One some models for developmental systems. Part VI: patterns in developmental systems, *International Journal of Systems Science*, vol. 17, pp. 559-567.

P.M.B. VITÁNYI, J. KARHUMÄKI, A. EHRENFEUCHT and G. ROZENBERG (1976), Growth functions *in* J. TOU (ed.), *Advances in Information System Science*, Plenum Press, New York, pp. 104-141.

S. WĘGRZYN, J.C. GILLE and P. VIDAL (1982), On some models for developmental systems. Part I: organism structure on the basis of elementary transformations, *International Journal of Systems Science*, vol. 13, pp. 359-378.

S. WĘGRZYN, J.C. GILLE and P. VIDAL (1984), Sur les propriétés des systèmes évolutifs engendrés par un mot créateur bouclé, *Revue française d'automatique, d'informatique et de recherche opérationnelle (Automatique)*, vol. 18, pp. 79-94.

S. WĘGRZYN, J.C. GILLE and P. VIDAL (1986), Clonage et principe de la récapitulation dans la théorie des systèmes évolutifs, *Bulletin of the Polish Academy of Sciences (Technical Sciences)*, vol. 34, pp. 103-109.

S. WĘGRZYN, P. VIDAL and J.C. GILLE (1981), Graphes et matrices des systèmes évolutifs, *Podstawy sterowania*, vol. 11, pp. 325-336.

S. WĘGRZYN, P. VIDAL and J.C. GILLE (1985), Quelques propriétés structurales des systèmes évolutifs et des organismes vivants, *Podstawy sterowania*, vol. 15, pp. 239-260.

S. WĘGRZYN, P. VIDAL and J.C. GILLE (1986), Mutations et théorie des systèmes évolutifs, *Bulletin of the Polish Academy of Sciences (Technical Sciences)*, vol. 34, pp. 121-124.

R. WINIARCZYK (1981), O funkcji wzrostu procesów rozwojowych [Growth function of developmental processes], *Podstawy sterowania*, vol. 11, pp. 313-324.

R. WINIARCZYK (1983), O syntezie słowa tworzącego w systemach ewolucyjnych [Constructing the generating word of a developmental system], *Podstawy sterowania*, vol. 13, pp. 65-74.

Chapter 3

SYNCHRONOUS DEVELOPMENTAL SYSTEMS WITH FEEDBACK

3.1 GENERAL

To the best of our knowledge the concept of feedback in developmental systems was first introduced by Lück and Lück, 1976 (see also Gille, Węgrzyn and Vidal, 1981).

A generating word has feedback if the first cell a is regenerated by a certain operation A_h

$$A_h\,(h) = a \qquad \text{or} \qquad A_h\,(h) = ah \qquad \text{or} \qquad A_h(a) = a(h) \qquad (3\text{-}1)$$

so that all the letters of such a generating word, including the first one, have an antecedent.

Feedback may be introduced from any letter except from a letter which was originally a B or a C:

(a) an S operation ($h \rightarrow h$) can become either a T operation ($h \rightarrow a$) or an L or R operation [$h \rightarrow ah$ or $h \rightarrow a(h)$];

(b) a T operation ($h \rightarrow i$) or an L or R operation [$h \rightarrow ih$ or $h \rightarrow i(h)$] can become a B or a C [$h \rightarrow ia$ or $h \rightarrow i(a)$].

In Figure 3-1: h is fed back to a, the operation $T\,(h \rightarrow i)$ becoming a $B\,(h \rightarrow ia)$. Other examples will be given later.

The generating graph of such a system possesses a *loop* around the A_hA_a path; it will be termed *circular*. If A_h was the last letter of the generating word before feedback was introduced, we shall speak of a *global loop*; if not, of *local loop*. In the condensed writing of the generating word the letter A_h of the operation from which the feedback starts will bear an F (for *f*eedback) superscript. — In contrast a generating word without feedback will be called *linear* or *rectilinear*.

For the four systems shown in Figure 3-2 (to be compared to Figure 2-25) the circular generating words are respectively:

$LCSRT^F$ and $LTTTL^F$ (global loops)

CLB^FSRS and $LBLLB^FLSS$ (local loops)

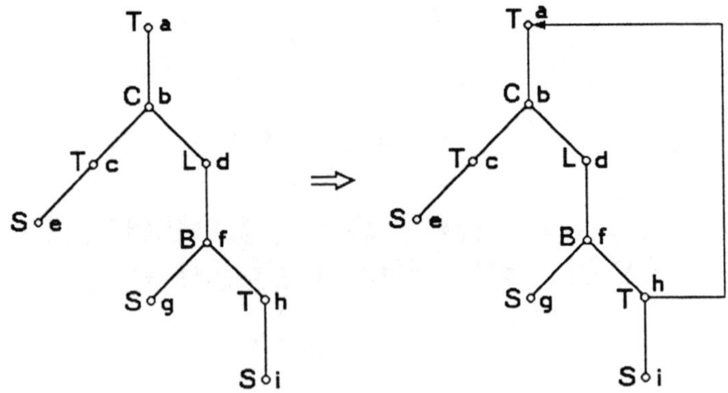

Fig. 3-1. Feedback (from *h* towards *a*) in a developmental system.

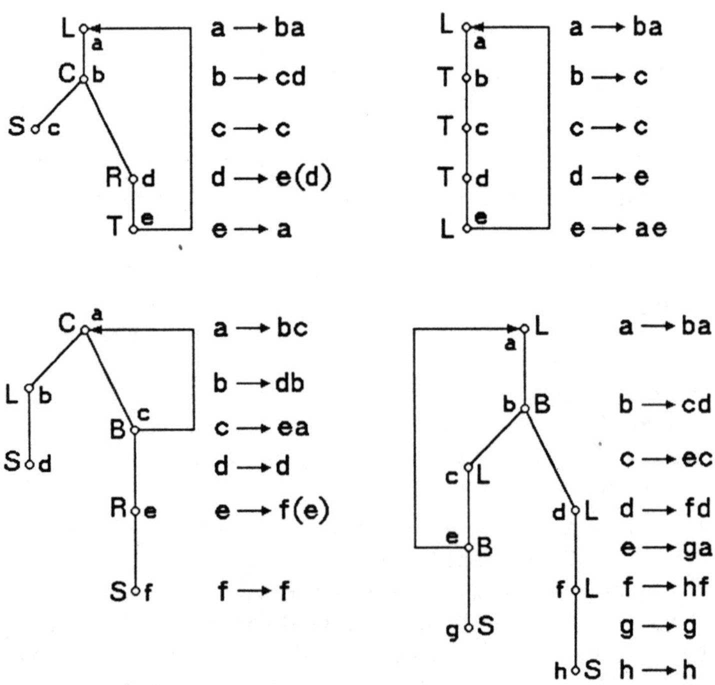

Fig. 3-2. Graph and elementary operations for the four generating graphs shown in Figure 2-26 after feedback has been introduced.

Note. — Feedback towards another operation than the first will not be considered here because it would violate our axiom that each cell should have only one antecedent.

For example, in the case of the system

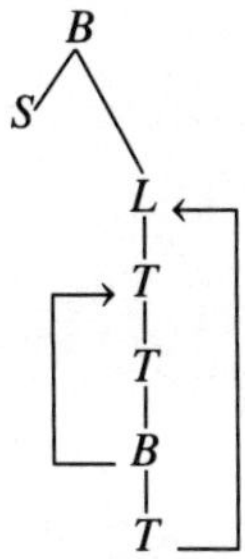

$$a \to b(c) \quad b \to b \quad c \to dc \quad d \to e \quad e \to f \quad f \to dg \quad g \to c$$

$$\mathbf{M} = \begin{bmatrix} 0 & 1 & 1 & 0 & 0 & 0 & 0 \\ 0 & 1 & 0 & 0 & 0 & 0 & 0 \\ 0 & 0 & 1 & 1 & 0 & 0 & 0 \\ 0 & 0 & 0 & 0 & 1 & 0 & 0 \\ 0 & 0 & 0 & 0 & 0 & 1 & 0 \\ 0 & 0 & 0 & 1 & 0 & 0 & 1 \\ 0 & 0 & 1 & 0 & 0 & 0 & 0 \end{bmatrix}$$

it is easily seen that a c and a d cell would have two antecedents (b and g; c and f, respectively). Note that the evolution matrix is not upper-triangular as a consequence of the presence of ones anywhere in the lower-left part (not only in the first column).

3.2 ANALYSIS AND SYNTHESIS

The evolution of developmental systems with feedback can be analyzed by the same numerical methods as the evolution of systems with a linear generating word. Reciprocally, generating words with feedback can be constructed in order to reproduce some predetermined patterns (either desired or found in natural organisms) (Węgrzyn, Gille and Vidal, 1982, p. 373-378).

3.2.1 Example I

Consider the system

$$DS_R = \frac{L\,C_{\pm 45}\,S\,T^F}{a,b,c,d} \tag{3-2}$$

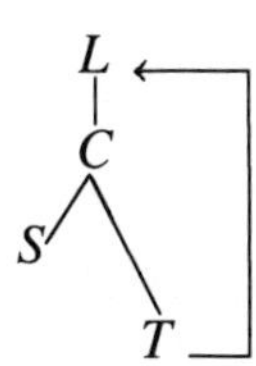

$$a \to ba \quad b \to c(d) \quad c \to c \quad d \to a \tag{3-3}$$

which differs from DS_C (§ 1.2.3) by the introduction of feedback from the second S (now a T) operation.

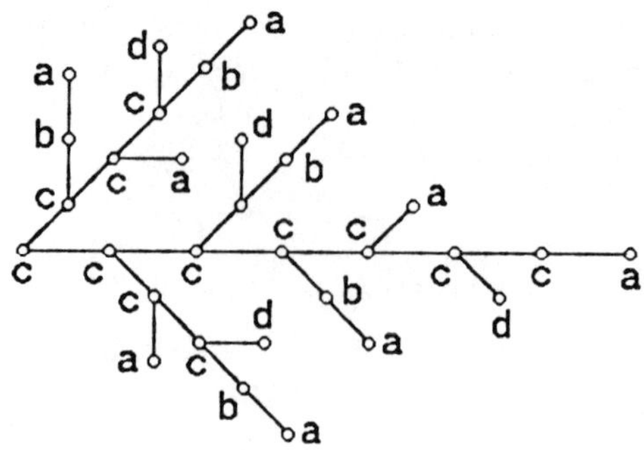

Fig. 3-3. $k = 7$ developmental stage of $DS_R = \dfrac{L\,C\,S\,T^F}{a,b,c,d}$.

If the initial condition is one a cell, the development proceeds as follows:

$$\left.\begin{array}{l} DS_R(0) = a \quad DS_R(1) = ba \quad DS_R(2) = c(d)ba \quad DS_R(3) = c(a)c(d)ba \\ DS_R(4) = c(ba)c(a)c(d)ba \qquad DS_R(5) = c(c(d)ba)c(ba)c(a)c(d)ba \\ DS_R(6) = c(c(a)c(d)ba)c(c(d)ba)c(ba)c(a)c(d)ba \\ DS_R(7) = c(c(ba)c(a)c(d)ba)c(c(a)c(d)ba)c(c(d)ba)c(ba)c(a)c(d)ba \end{array}\right\} \quad (3\text{-}4)$$

The $k = 7$ developmental stage is shown in Figure 3-3.

Comparison with Figure 1-3 and equations 1-8, 1-9 shows that:

1) the generated "tree" has branches and subbranches, although the generating word contains only one C operation;

2) the growth is faster than before feedback was present:

$$V(4) = 9 \qquad V(5) = 14 \qquad V(6) = 21 \qquad V(7) = 31$$
$$V(8) = 46 \qquad V(9) = 68$$

(No analytidal expression for $V(k)$ is apparent. See Paragraphs 3.4.1 *in fine* and 3.4.3 B.2 *in fine*.)

Note in passing the resemblance between Figures 3-3 and 2-22. However the number of cell categories is smaller in the present case.

3.2.2 Example II

Consider the developmental system

$$DS_S = \frac{L\,T\,T\,T\,B\,S\,C^F\,S}{a,b,c,d,e,f,g,h} \qquad (3\text{-}5)$$

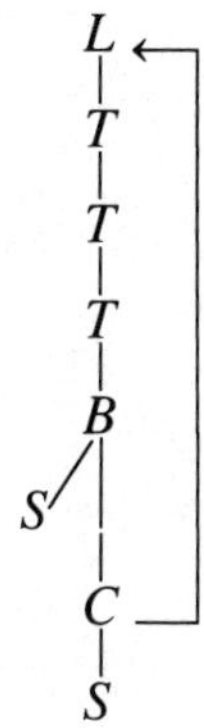

$$\begin{aligned} a &\to ba \\ b &\to c \\ c &\to d \\ d &\to e \\ e &\to fg \\ f &\to f \\ g &\to h(a) \\ h &\to h \end{aligned} \qquad (3\text{-}6)$$

The first developmental stages are one-dimensional (rectilinear) filaments:

$$DS_S(0) = a \qquad DS_S(1) = ba \qquad DS_S(2) = cba$$

$$DS_S(3) = dcba \qquad DS_S(4) = edcba \qquad DS_S(5) = fgedcba$$

With $DS_S(6)$ the feedback $[g \to h(a)]$ sets in and starts creating branches:

$$DS_S(6) = fh(a)fgedcba \qquad DS_S(7) = fh(ba)fh(a)fgedcba$$

From $DS_S(12)$ on, subbranches are generated by the C^F operation acting on the g cells of the branches:

$$DS_S(12) = fh(fh(a)fgedcba)fh(fgedcba)fh(edcba)fh(dcba)fh(cba) \\ fh(ba)fh(a)fgedcba$$

$$DS_S(13) = fh(fh(ha)fh(a)fgedcba)fh(fh(a)fgedcba)fh(fgedcba)fh(edcba) \\ fh(dcba)fh(cba)fh(ba)fh(a)fgedcba$$

The interest of this system lies in the fact that Lindenmayer (see Herman and Rozenberg, 1975, p. 31-35) used it as a model for the vegetative development of *Syringa vulgaris*. Our $DS_S(13)$ pattern (Fig. 3-4) is identical to the one published by him.

3.2.3 Example III

It is possible to generate a developmental system DS_T in which the first three branches have no subbranches but the subsequent branches evince indefinite-order subdivision. We shall do so by means of the following word with feedback

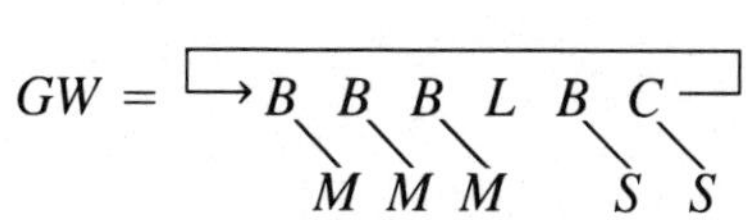

(3-8)

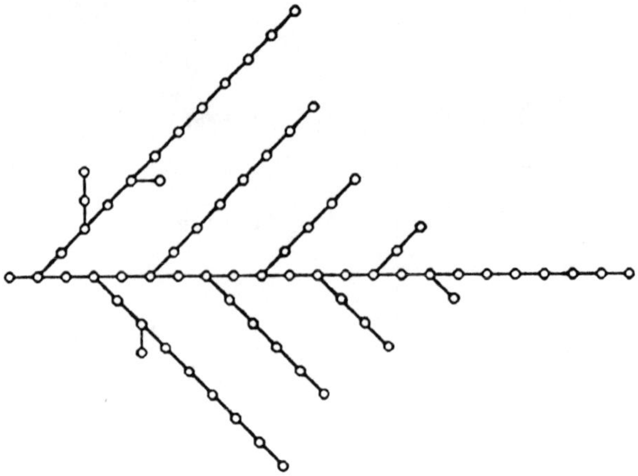

Fig. 3-4. $k = 13$ developmental stage of system DS_S (eq. 3-6).

in which the letter M stands for a subword defined by

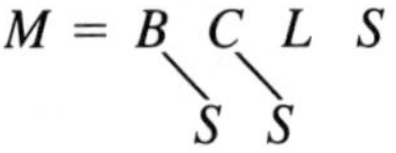

i.e. by the following sequence of cell transformation

$$\begin{array}{l} m \rightarrow sg \\ g \rightarrow s(h) \\ h \rightarrow hs \\ s \rightarrow s \end{array}$$

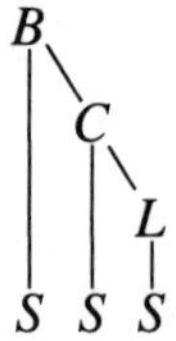

The elementary cell transformations of the overall system are

$$\begin{array}{l} a \rightarrow mb \\ b \rightarrow mc \\ c \rightarrow md \\ d \rightarrow ed \\ e \rightarrow sf \\ f \rightarrow s(a) \end{array} \qquad (3\text{-}9)$$

The sequence of states

$$\left.\begin{array}{l} DS_T(0) = \mathrm{a} \qquad DS_T(1) = mb \qquad DS_T(2) = sgmc \\ DS_T(3) = ss(h)sgmd \qquad DS_T(4) = ss(sh)ss(h)sged \\ DS_T(5) = ss(ssh)ss(sh)sfed \\ DS_T(6) = ss(sssh)ss(ssh)ss(sh)ss(a)sfed \qquad [(a) = \mathrm{loop}] \end{array}\right\} \tag{3-10}$$

is generated according to

Procedure $P(n)$

$n = 0,\ a$
$n = 1,\ mb$
$n = 2,\ sgmc$
$n = 3,\ ss(h)sgmd$
$n = 4,\ ss(sh)ss(h)sged$
$n = 5,\ ss(ssh)ss(sh)ss(h)sfed$
$n > 5,\ ss(A(n-3))ss(A(n-4))ss(A(n-5)ssB(n-6)sfed$

Procedure $A(n)$

$n = 1,\ sh$
$n > sA(n-1)$

Procedure $B(n)$

$n = 0,\ (P(1))$
$n > 0,\ (P(n))ssB(n-1)$

Figure 3-5 shows the structure of $DS_T(13)$.

Note that Rozenberg (see Herman and Rozenberg, 1975, p. 43-52) accounted for the development of *Callithamnion Roseum* by constructing the model based on the following elementary cell operations

$$\left.\begin{array}{ll} d \to ed & e \to sf \\ f \to s(a) & s \to s \\ a \to mb & m \to sg \\ b \to mc & g \to s(h) \\ c \to md & h \to sh \end{array}\right\} \tag{3-11}$$

which model is equivalent to ours. The structure of Figure 3-5 is identical to that published by him.

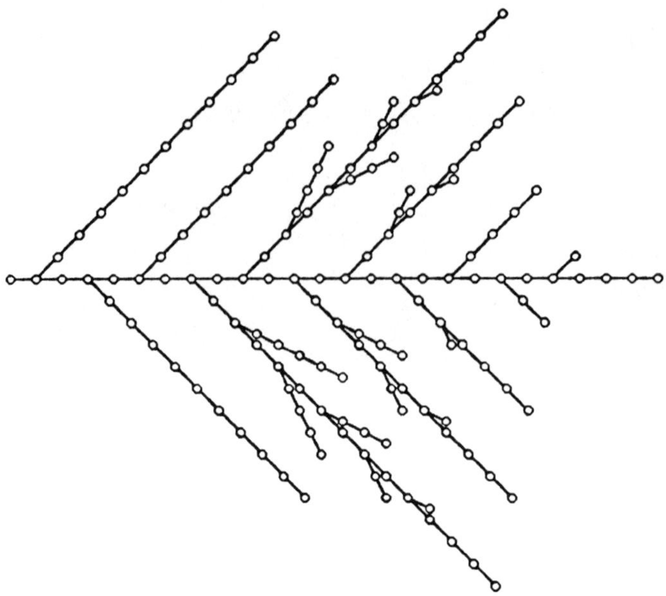

Fig. 3-5. $k = 13$ developmental stage of system DS_T (eq. 3-9).

3.3 STRUCTURAL PROPERTIES

It will now be seen that developmental systems with feedback in their generating word have structural properties that differ from those of developmental systems without feedback. Essentially, the concept of hierarchy does not apply any more; systems with feedback are apt to total regeneration after quasi total destruction and are extremely sensitive to mutation.

3.3.1 Hierarchy, Patterns

The existence of hierarchy inside developmental systems without feedback (§ 2.2.1) is a consequence of the fact that the A_i operations are executed in the natural order of their indexes, which is expressed by the binary-tree shape of the generating graph. If the generating word has feedback from a letter A_h towards the first letter A_a, the principle of hierarchy does not hold any more because of the interference of the recurrently renewed generation of the subword $A_a \ldots A_h$.

Therefore subsystems are more difficult to identify. For example, by comparing Figures 2-31 and 3-3 the reader will notice that the cellular composition of the branches of the latter does not suggest the idea of an organ composed of definite tissues as do the branches of the former.

It is possible, however, to identify patterns inside some classes of developmental systems with feedback. For this subject the reader is referred to our articles: Węgrzyn, Vidal and Gille, 1984; Węgrzyn, Gille and Vidal, 1986.

3.3.2 Cloning, Grafting

The possibility of "constructing" complex developmental systems by combining two generating words exists also in the case of circular generating words. Recall that it consists in placing the "root" of one graph at an external node of the other: such a recombination is the image of the procedure of cloning (see Paragraph 2.2.3).

Consider the developmental system (see Paragraph 1.2.4)

$$DS_{D2} = \frac{GW_{D2}}{Z_{D2}} = \frac{L\,C^F\,S}{a,b,c} \tag{1-14}$$

and suppose that the $P = LCSS$ fragment of another generating word extended over the $Z = \{c,d,e,f\}$ cell set is introduced in GW_{D2} at the last S. The result of this cloning procedure in the more complex system

$$DS_{D3} = \frac{L\,C^F\,P}{a,b,Z} = \frac{L\,C^F\,L\,C\,S\,S}{a,b,c,d,e,f}$$

The $k = 5$ developmental stages are respectively

$$DS_{D2}(5) = ba(c)a(c)baca(c)ba(c)ba(c)a(c)ba$$

$$DS_{D3}(5) = ba(dc)a(c)ba(e(f)e(f)dc)a(c)ba(e(f)dc)ba(dc)a(c)ba$$

They are shown in Figure 3-6.

The difference between cloning and *grafting* should be emphasized. Grafting consists of extracting a fragment of an organism and inserting it into another organism.

If for example the c cell for the $k = 2$ stage of the above DS_{D2} system $[DS_{D2}(2) = a(c)ba]$ is replaced by the initial cell d of the developmental system (adapted from DS_F, eq. 2-1)

$$DS_{D4} = \frac{L\,C\,S\,S}{d,e,f,g}$$

a new developmental system is obtained

$$DS_{D5} = \frac{L\,C^F\,S\,L\,C\,S\,S}{a,b,c,d,e,f,g}$$

with the initial condition

$$DS_{D5}(0) = a(d)ba$$

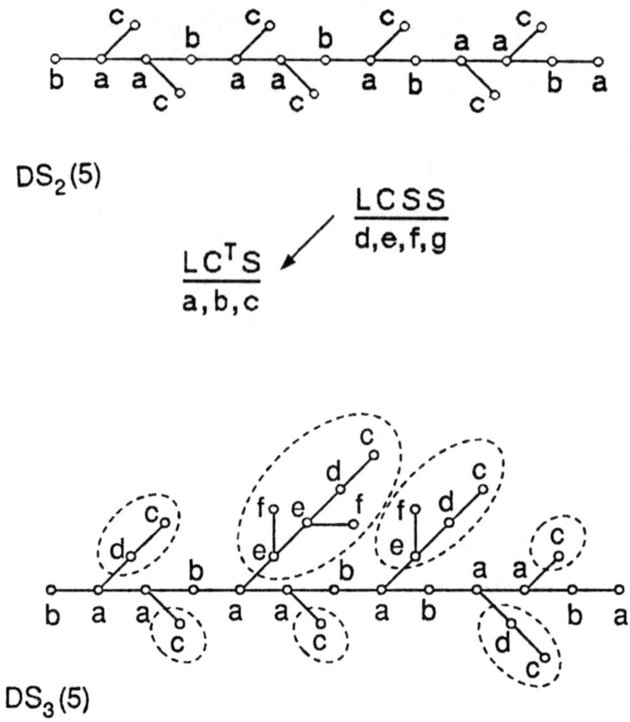

Fig. 3-6. $k = 5$ developmental stage of DS_{D2} system before and after cloning.

The $k = 3$ developmental stage

$$DS_{D5}(3) = ba(f(g)ed)a(c)ba(c)a(c)ba(c)ba(c)a(c)ba$$

corresponds to the $k = 5$ developmental stage of the original system DS_{D2} (prior to grafting): see Figure 3-7.

3.3.3 Influence of initial conditions; regeneration

Consider the developmental system the generating word of which is shown in Figure 3-8. Suppose the initial state is one a cell: the $k = 1,2,3$ stages are the same as if the generating word were the rectilinear word $A_aA_bA_cA_dA_eS$ but at the $k = 6$ stage the operation A_f $(f \to a)$ reintroduces the initial cell a, from which a new development starts. The same holds if the initial condition consists of a, b, c, f cell(s), i.e., of cells subjected to

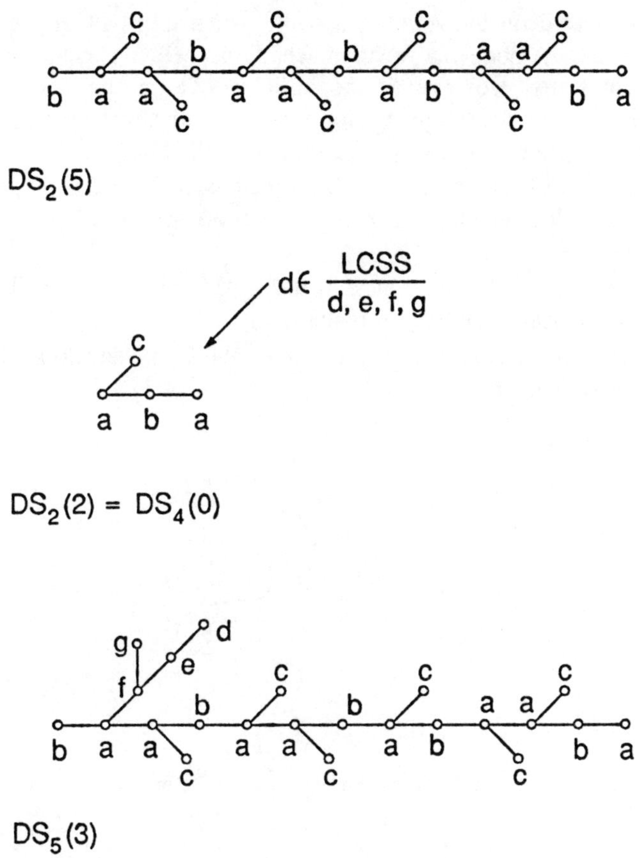

Fig. 3-7. $k = 5$ developmental stage of DS_{D2} system before and after grafting.

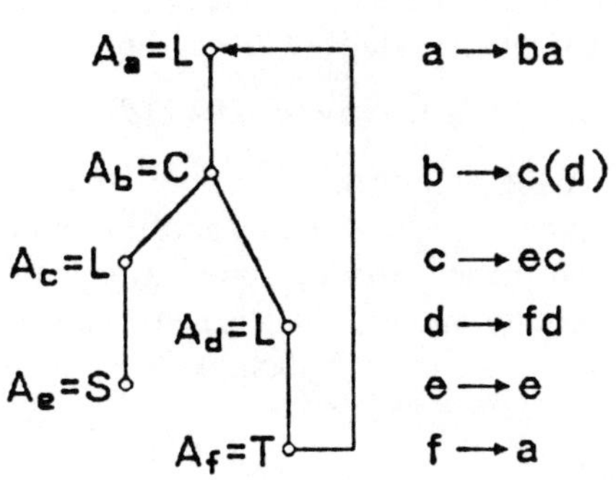

Fig. 3-8. System with feedback (circular generating graph).

operations located on the $A_aA_bA_dA_f$ path, which is a part of the feedback loop: once the A_f node has been reached the cell a appears and a new development starts. However if the initial condition consists of cell(s) c associated with the operation A_c, the node A_f will never be reached: only the pattern generated by the subword A_cA_e will develop.

In other words the structure developed is *independent of the initial condition provided* the latter consists of cell(s) associated to (an) operation(s) lying *on the feedback loop*.

As a consequence if the system is destroyed but one such cell survives, *complete regeneration* will take place.

Consider e.g. the system DS_R (eq. 3-2), the $k = 7$ developmental stage of which is shown in Figure 3-3.

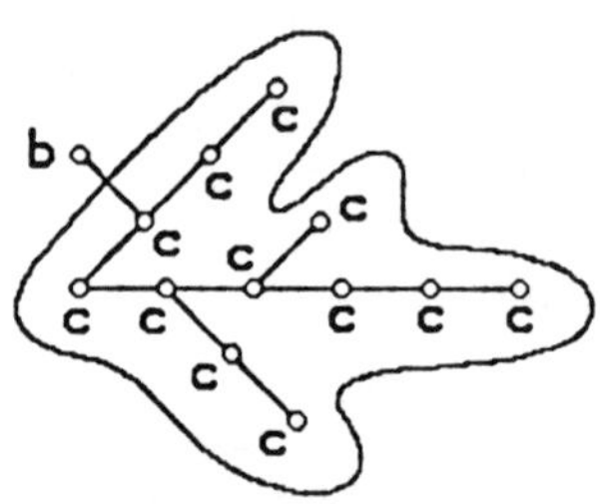

Fig. 3-9. Partial destruction of system DS_R (Fig. 3-3): only the stagnant cells and one b cell survive.

Suppose now that, as a consequence of some external circumstances, all the cells have been destroyed except (see Fig. 3-9) the stagnant cells (outlined in the figure) and one cell of category b. Regeneration will occur and the $k = 9$ developmental stage will be (Fig. 3-10)

$$DS_R(9) = cc(c(ba)c(a)c(d)ba)c(c(a)c(d)ba)c(c(d)ba)c(ba)c(a)c(d)ba$$

It is known that the ability to regenerate completely after important, even quasi total destruction, is characteristic of lower organisms. (Accounting in a more detailed manner for the complex phenomena of regeneration in biology would require more sophisticated mathematical models, e.g. models including internal feedback loops.)

3.3.4 Mutation

The consequences of changing one letter in the generating word are much more drastic if the latter has feedback than in the rectilinear case.

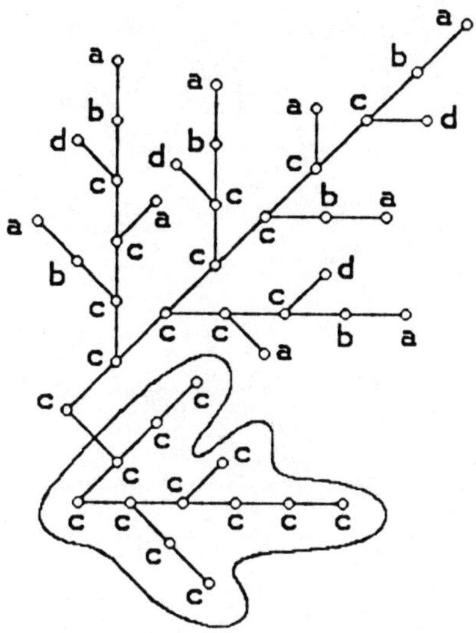

Fig. 3-10. Regeneration of system with feedback DS_R after the partial destruction shown in Figure 3-9.

Consider for example the developmental system the generating word of which is

$$\begin{array}{ccc} T \leftarrow\!\urcorner & & a \to b \\ | \quad\;\; | & & \\ C \;\lrcorner & & b \to c(a) \\ | & & \\ S & & c \to c \end{array} \qquad (3\text{-}12)$$

and suppose the T operation has become an L, so that the "mutant" is DS_{D2}

$$(1\text{-}14) \qquad \begin{array}{ccc} L \leftarrow\!\urcorner & & a \to \mathbf{ba} \text{ (modified operation)} \\ | \quad\;\; | & & \\ C \;\lrcorner & & b \to c(a) \\ | & & \\ S & & c \to c \end{array} \qquad (3\text{-}13)$$

If the initial condition is one a cell, the $k = 6$ developmental stage consists, for the initial system, of one spiral chain (Fig. 3-11 left)

$$c(c(c(a)))$$

but for the modified system it consists of a complete tree with branches and subbranches (see end of Paragraph 1.2.4):

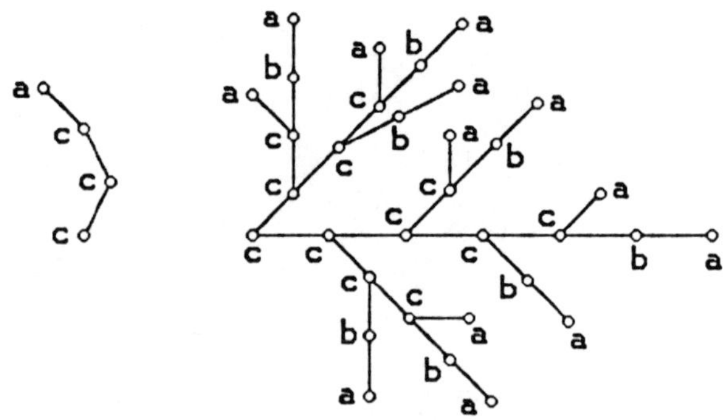

Fig. 3-11. System with feedback $\dfrac{T\,C^F\,S}{a,b,c}$ at $k = 6$ stage before (left) and after (right) the substitution (mutation) $T \rightarrow L$ *inside* the loop. Note the complete change of structure.

$$DS_{D2}(6) = c(c(c(a)ba)c(ba)c(a)ba)c(c(ba)c(a)ba)c(c(a)ba)c(ba)c(a)ba$$

(Fig. 3-11 right).

It is thus observed that the *structure* of systems generated by a generating word with feedback can be *entirely altered* by one change of letter in the generating word, provided the change affects an operation lying on the feedback loop.

In other words, developmental systems with feedback are much more *sensitive to changes* in the generating word than are systems without feedback.

Note that the complete change of structure observed in Figure 3-11 has occurred because the altered operation lies inside the feedback loop. In fact, altering an operation lying outside the loop causes modifications of the system but does not alter its structure. Figure 3-12 show the two developmental systems

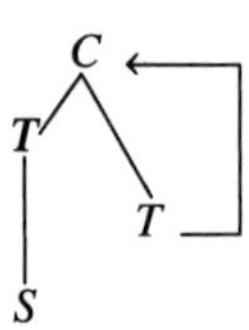

$a \rightarrow b(c)$
$b \rightarrow \mathbf{d}$
$c \rightarrow a$
$d \rightarrow d$

⇛

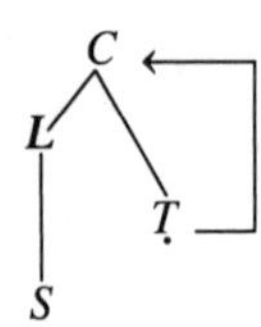

$a \rightarrow b(c)$
$b \rightarrow \mathbf{db}$
$c \rightarrow a$
$d \rightarrow d$

at their $k = 7$ developmental stage: respectively

$$d(d(d(b(c)))) \quad \text{and} \quad ddddddb(ddddb(ddb(d(c))))$$

It is observed that only the relative size of the segments of the system, not its structure, has been altered.

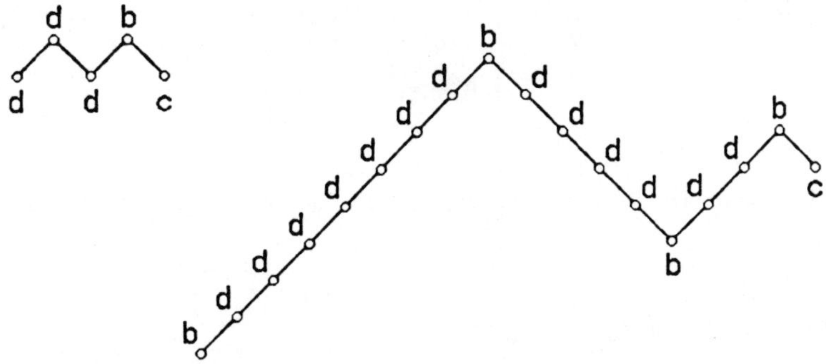

Fig. 3-12. Effect of mutation ($T \to L$ *outside* the loop) on a developmental system with feedback. The structure remains unaltered.

3.4 QUANTITATIVE GROWTH

It was shown in Section 2.3 that developmental systems with a linear generating word grow at a polynomial rate at a power which primarily depends on the number of active (L and R) operations. It will now be seen that developmental systems with feedback generally grow exponentially, the number of active operations still playing a fundamental role. (References: Gille, Vidal and Węgrzyn, 1985; Węgrzyn, Gille and Vidal, 1985.)

3.4.1 Numerical approach

The evolution of the cellular composition and of the size can be analyzed by step by step computing $a(k)$, $b(k)$, ..., $n(k)$. But litteral analytical expressions for $\mathbf{Y}(k)$ and $V(k)$ can be found only in exceptional cases.

Example 1: $DS_U = \dfrac{L\,L\,L^F}{a,b,c}$

$$\begin{array}{ccc} L \leftarrow\!\urcorner & \quad & a \to ba \\ | & & \\ L & & b \to cb \\ | & & \\ L \;\lrcorner & & c \to ac \end{array} \qquad (3\text{-}14)$$

The successive stages are:

$$a \quad ba \quad cbba \quad accbcbba \quad baacaccbaccbcbba \quad \ldots \qquad (3\text{-}15)$$

At each step each cell generates two cells; therefore the size is multiplied by 2. Hence

$$V(k) = 2^k \tag{3-16}$$

Example II: $DS_V = \dfrac{L\,T^F}{a,b}$

By incorporating feedback into DS_A (§ 1.2.1) the following system is obtained

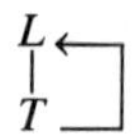

$$a \to ba \qquad b \to a \tag{3-17}$$

The successive stages are

$$\left.\begin{matrix} a \quad ba \quad aba \quad baaba \quad ababaaba \quad baabaababaaba \\ ababaababaabaababaaba \quad baabaababaabaababaababaabaababaaba \quad \ldots \end{matrix}\right\} \tag{3-18}$$

The sequences

$$a(k)\colon 1, 1, 2, 3, 5, 8, 13, 21, 34, 55, \ldots$$

$$b(k)\colon 0, 1, 1, 2, 3, 5, 8, 13, 21, 34, \ldots$$

are obviously *Fibonacci sequences*(1):

$$a(k + 1) = a(k) + a(k - 1) \qquad b(k + 1) = b(k) + b(k - 1)$$

$$V(k + 1) = V(k) + V(k - 1)$$

This fact, first observed by Szilard (1971, p. 37), is a consequence of the recursive equations

$$a(k + 1) = a(k) + b(k) \qquad b(k + 1) = a(k)$$

Example III: $DS_W = \dfrac{L\,T\,T \ldots T\,T^F}{a,b,c,\ldots,t,n}$

More generally, consider the following generating graph, consisting of one L and of (n − 1) T, with feedback from the last of them:

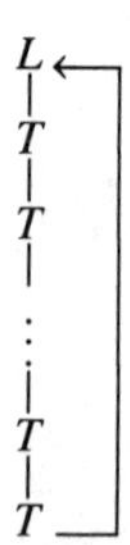

$$a \to ba \qquad b \to c \qquad c \to d \qquad \ldots \qquad t \to u \qquad u \to a \tag{3-19}$$

already investigated by the present authors (Węgrzyn, Gille and Vidal, 1982, pp. 373-374).

(1) The importance of Fibonacci sequences in phyllotaxy has been pointed out by Roger Jean (1978, pp. 41-_3).

The first n developmental stages ($k = 0, 1, \ldots, n-1$) are

$$a \quad ba \quad cba \quad \ldots \quad tsr \quad \ldots \quad cba \quad utsr \quad \ldots \quad cba$$

then the T^F $(u \rightarrow a)$ operation starts acting, whence

$$aut \quad \ldots \quad cba \quad baut \quad \ldots \quad cba \quad cbaut \quad \ldots$$
$$cba \quad dcbaut \quad \ldots \quad cba$$

The cellular composition is thus

$$\left.\begin{array}{rl} \mathbf{Y}(0) = [1 & 0 \quad 0 \quad \ldots \quad 0 \quad 0 \quad 0] \\ \mathbf{Y}(1) = [1 & 1 \quad 0 \quad \ldots \quad 0 \quad 0 \quad 0] \\ \mathbf{Y}(2) = [1 & 1 \quad 1 \quad \ldots \quad 0 \quad 0 \quad 0] \\ & \ldots \\ \mathbf{Y}(n-1) = [1 & 1 \quad 1 \quad \ldots \quad 1 \quad 1 \quad 1] \end{array}\right\} \tag{3-20a}$$

and for $k \geqslant n$

$$\left.\begin{array}{rl} \mathbf{Y}(n) = [2 & 1 \quad 1 \quad \ldots \quad 1 \quad 1 \quad 1] \\ \mathbf{Y}(n+1) = [3 & 2 \quad 1 \quad \ldots \quad 1 \quad 1 \quad 1] \\ \mathbf{Y}(n+2) = [4 & 3 \quad 2 \quad \ldots \quad 1 \quad 1 \quad 1] \end{array}\right\} \tag{3-20b}$$

It is observed that

$$\mathbf{Y}(n) = \mathbf{Y}(0) + \mathbf{Y}(n-1)$$

and more generally

$$\mathbf{Y}(k+n) = \mathbf{Y}(k+n-1) + \mathbf{Y}(k) \tag{3-21}$$

In other words, all the $i(k)$ are *generalized Fibonacci sequences*, i.e. solutions of

$$x(k+n) = x(k+n-1) + x(k) \tag{3-22}$$

with

$$x(0) = x(1) = \ldots = x(n-1) = 1$$

Example IV: $DS_R = \dfrac{L\,C\,S\,T^F}{a,b,c,d}$

The successive developmental stages have been computed in Paragraph 3.2.1 (eq. 3-4). The evolution of the cellular composition is:

$\mathbf{Y}(0) = [1$	0	0	0]	$V(0) = 1$	$V(0) - c(0) = 1$
$\mathbf{Y}(1) = [1$	1	0	0]	$V(1) = 2$	$V(1) - c(1) = 2$
$\mathbf{Y}(2) = [1$	1	1	1]	$V(2) = 4$	$V(2) - c(2) = 3$
$\mathbf{Y}(3) = [2$	1	2	1]	$V(3) = 6$	$V(3) - c(3) = 4$
$\mathbf{Y}(4) = [3$	2	3	1]	$V(4) = 9$	$V(4) - c(4) = 6$
$\mathbf{Y}(5) = [4$	3	5	2]	$V(5) = 14$	$V(5) - c(5) = 9$
$\mathbf{Y}(6) = [6$	4	8	3]	$V(6) = 21$	$V(6) - c(6) = 13$

$$\mathbf{Y}(7) = [9 \quad 6 \quad 12 \quad 4] \qquad V(7) = 31 \qquad V(7) - c(7) = 19$$

$$\mathbf{Y}(8) = [13 \quad 9 \quad 18 \quad 6] \qquad V(8) = 46 \qquad V(8) - c(8) = 28$$

$$\mathbf{Y}(9) = [19 \quad 13 \quad 27 \quad 9] \qquad V(9) = 68 \qquad V(9) - c(9) = 41$$

It is seen that $a(k)$, $b(k)$ and $d(k)$ — the numbers of the cells acted upon by the operations (L,C,T) lying inside the loop — are *generalized Fibonacci sequences*

$$j(k+3) = j(k+2) + j(k) \qquad j = a,b,d$$

Such is also their sum, which is

$$a(k) + b(k) + d(k) = V(k) - c(k)$$

(But $V(k)$ is not a Fibonacci sequence, because of the $c(k)$ term.)

3.4.2 Matrix approach

When the generating graph possesses a loop, the evolution matrix is *not upper-triangular any more:* the feedback from the A_h operation is expressed by the presence of a one at the intersection of the h-th row (the last row if the loop is a global loop) and the first column.

The matrices of the four systems shown in Figure 3-2 are:

$$\mathbf{M}(LCSRT^F) = \begin{bmatrix} 1 & 1 & 0 & 0 & 0 \\ 0 & 0 & 1 & 1 & 0 \\ 0 & 0 & 1 & 0 & 0 \\ 0 & 0 & 0 & 1 & 1 \\ 1 & 0 & 0 & 0 & 0 \end{bmatrix} \qquad \mathbf{M}(LTTTL^F) = \begin{bmatrix} 1 & 1 & 0 & 0 & 0 \\ 0 & 0 & 1 & 0 & 0 \\ 0 & 0 & 0 & 1 & 0 \\ 0 & 0 & 0 & 0 & 1 \\ 1 & 0 & 0 & 0 & 1 \end{bmatrix}$$

$$\mathbf{M}(CLB^FSRS) = \begin{bmatrix} 0 & 1 & 1 & 0 & 0 & 0 \\ 0 & 1 & 0 & 1 & 0 & 0 \\ 1 & 0 & 0 & 0 & 1 & 0 \\ 0 & 0 & 0 & 1 & 0 & 0 \\ 0 & 0 & 0 & 0 & 1 & 1 \\ 0 & 0 & 0 & 0 & 0 & 1 \end{bmatrix} \qquad \mathbf{M}(LBLLB^FLSS) = \begin{bmatrix} 1 & 1 & 0 & 0 & 0 & 0 & 0 & 0 \\ 0 & 0 & 1 & 1 & 0 & 0 & 0 & 0 \\ 0 & 0 & 1 & 0 & 1 & 0 & 0 & 0 \\ 0 & 0 & 0 & 1 & 0 & 1 & 0 & 0 \\ 1 & 0 & 0 & 0 & 0 & 0 & 1 & 0 \\ 0 & 0 & 0 & 0 & 0 & 1 & 0 & 1 \\ 0 & 0 & 0 & 0 & 0 & 0 & 1 & 0 \\ 0 & 0 & 0 & 0 & 0 & 0 & 0 & 1 \end{bmatrix}$$

The fundamental relation

$$\text{(2-51)} \qquad \mathbf{Y}(k) = \mathbf{Y}(0)\mathbf{M}^k$$

still holds. But, since $\mathbf{M}$ is not triangular any more, its characteristic values $\lambda_1, \ldots, \lambda_n$ are no longer 0 and 1. Therefore the growth occurs no more at a polynomial rate (as was the case for systems without feedback), but *exponentially* as

$$\lambda_{\text{sup}}^k$$

where the "dominant mode" is

$$\lambda_{\text{sup}} = \sup(|\lambda_1|, |\lambda_2|, \ldots, |\lambda_n|)$$

Some information on λ_{sup} can be obtained by means of Gershgorin's theorem (for example, Legras 1963, pp. 58-61 or Marcus and Minc 1964, pp. 145-150) or by means or Frobenius's theorem on non-negative matrices (for example, Gantmacher 1959, pp. 53-54, 64-65).

In fact: (1) all the elements of **M** that lie in the main diagonal are zeros and ones, and (2) each row is made up of zeros and of one or two (not more) ones.

Thus from Gershgorin's theorem

$$|\lambda_i - 0| \leq 2 \qquad |\lambda_i - 1| \leq 1 \qquad i = 1, \ldots, n$$

whence

$$\lambda_{\text{sup}} \leq 2$$

As a consequence of Frobenius' theorem the characteristic value which has the greatest absolute value is real, positive, has simple order of multiplicity and satisfies

$$1 \leq \lambda_{\text{sup}} \leq 2$$

In conclusion, the rate of *growth is not faster than* 2^k.

This result, which was first established by Paz and Salomaa (1973, p. 333), is not surprising, since at any stage of development one cell can at most generate two cells (according to the definition of the elementary operations that constitute the generating word).

Altogether the easiest method for investigating the growth of developmental systems with feedback consists of resorting to the z (or to the discrete Carson) transform, which will be done in the next paragraph.

3.4.3 Transform approach

A) Theory

The relation that exists between the z-transforms of the sequences $\{i(k)\}$ and $\{j(k)\}$ of two cell categories which immediately follow each other on the generating graph was established in Paragraph 2.3.4:

$$j(z) = D_j(z)i(z) \tag{2-58}$$

where

$$D_j(z) = \frac{1}{z} \quad \text{or} \quad D_j(z) = \frac{1}{z-1} \tag{2-59}$$

according to whether the operation is T,B,C or is L,R,S.

Hence the fundamental relation

$$j(z) = a(z) \prod_{\text{path}:a}^{j} D_i(z) \tag{2-61}$$

still holds.

Recall that the Π on the right-hand side is the product of the z-transfer-functions $D_i(z)$ pertaining to the operations found on the path leading from a to j on the generating graph, the operation $D_a(z)$ being excluded.

For the systems shown in Figure 3-2: equations (2-62), (2-63) and (2-64) still hold, but for the fourth system

$$\prod_{\text{path}:a}^{e} D_i(z) = D_b(z)D_c(z)D_e(z) = \frac{1}{z}\frac{1}{z-1}\frac{1}{z} = \frac{1}{z^2}\frac{1}{z-1} \qquad [B,L,B]$$

should be written instead of (2-65).

The difference with the case of a linear generating word lies in the fact that $a(z)$ is no longer given by

$$a_0(z) = 1 \quad \text{or} \quad a_0(z) = \frac{z}{z-1} \tag{2-57}$$

according to whether A_a is T,B,C or is L,R, but is given by (see proof below)

$$a(z) = \frac{a_0(z)}{1 - F(z)} \tag{3-23}$$

where $F(z)$ is the *product of the z-transfer-functions* $D_i(z)$ of the operations *met on the loop path*, $D_a(z)$ (eq. 2-60) being *included*.

If u is the number of T,B,C operation and v the number of L,R operations located on the loop path, then

$$F(z) = \frac{1}{z^u}\frac{1}{(z-1)^v} \tag{3-24}$$

For the four systems shown in Figure 3-2, respectively:

$$F_1(z) = \frac{1}{z-1}\frac{1}{z}\frac{1}{z-1}\frac{1}{z} = \frac{1}{z^2(z-1)^2} \qquad [LCRT\text{: } u = 2,\ v = 2]$$

$$F_2(z) = \frac{1}{z-1}\frac{1}{z}\frac{1}{z}\frac{1}{z}\frac{1}{z-1} = \frac{1}{z^3(z-1)^2} \qquad [LTTTL\text{: } u = 3,\ v = 2]$$

$$F_3(z) = \frac{1}{z}\frac{1}{z} = \frac{1}{z^2} \qquad [CB\text{: } u = 2,\ v = 0]$$

$$F_4(z) = \frac{1}{z-1}\frac{1}{z}\frac{1}{z-1}\frac{1}{z} = \frac{1}{z^2(z-1)^2} \qquad [LBLB\text{: } u = 2,\ v = 2]$$

Combining (2-61) and (3-23) finally yields the z-transform of any $j(k)$ $(j = b,c,\ldots,n)$:

$$\boxed{j(z) = \frac{a_0(z)}{1 - F(z)} \prod_{\text{path}:a}^{j} D_i(z)} \tag{3-25}$$

Proof of equation 3-23.

If feedback is introduced, the difference equation governing $a(k)$ is

$$a(k + 1) = h(k)$$

in the first operation in the generating word is T, B or C, and is

$$a(k + 1) = a(k) + h(k)$$

if it is L or R.

Taking the z-transform, remembering the initial condition $a(0) = 1$ and taking equation (2-61) into account yields

$$za(z) - z = a(z) \prod_{\text{path}:a}^{h} D_i(z) \tag{3-26a}$$

$$za(z) - z = a(z) + a(z) \prod_{\text{path}:a}^{h} D_i(z) \tag{3-26b}$$

in the second case.

In these expressions the last factor is the product of the z-transfer-functions of the operations met on the loop path, $D_a(z)$ being excluded. In other words

$$\prod_{\text{path}:a}^{h} D_i(z) = \frac{F(z)}{D_a(z)} \tag{3-27}$$

Solving (3-25) with respect to $a(z)$ and taking (3-26) into account yields

$$a(z) = \frac{z}{z - \dfrac{F(z)}{D_a(z)}} \tag{3-28a}$$

in the first case, and

$$a(z) = \frac{z}{z - 1 - \dfrac{F(z)}{D_a(z)}} \tag{3-28b}$$

in the second case.

Expressions (3-28) can be brought to a unique form by multiplying their numerators and denominators by $D_a(z)$, i.e. by $1/z$ (according to equation 2-60a) in the first case

$$a(z) = \frac{1}{1 - F(z)}$$

and by $1/(z-1)$ (according to equation 2-60b) in the second case

$$a(z) = \frac{\dfrac{z}{z-1}}{1 - F(z)}$$

Both expressions are equivalent (see equations 2-57) to

$$a(z) = \frac{a_0(z)}{1 - F(z)} \tag{3-23}$$

B) Consequence

Expression (3-25) gives the explicit z-transforms of any $j(k)$ $(j = a,b,\ldots,n)$. In theory the sequences $j(k)$ themselves can be computed by performing the inverse transformation; but for systems of high order of complexity this leads to considerable work. Fortunately, essential information on $j(k)$ is obtained from the knowledge of the *poles* of $j(z)$.

Explicitly from (3-23) and (3-24)

$$a(z) = \frac{a_0(z)}{1 - \dfrac{1}{z^u(z-1)^v}} = \frac{z^u(z-1)^v}{z^u(z-1)^v - 1}\, a_0(z) \tag{3-29}$$

and from (3-25) for any $j = b,c,\ldots,n$

$$j(z) = \frac{z^u(z-1)^v}{z^u(z-1)^v - 1}\, a_0(z) \prod_{\text{path}:a}^{j} D_i(z) \tag{3-30}$$

It is thus seen that, as a consequence of the presence of feedback, the poles are no longer only 0 and 1: also present as poles are the roots of

$$z^u(z-1)^v - 1 = 0 \tag{3-31}$$

The consequence is that the *growth* does not occur at a polynomial rate any more, but (in the general case) at an *exponential rate* depending on the root of (3-31) which has the greatest absolute value z_{sup}, i.e. occurs as

$$z_{\text{sup}}^k \tag{3-32}$$

Several methods exist for numerically computing z_{sup} (for example, Démidovitch and Maron 1973, p. 424-431). The fact that, as a consequence of Frobenius's theorem on non-negative matrices (for example, Gantmacher 1959, p. 53-54, 64-65), the root of (3-31) with greatest absolute value is always real and positive, makes it possible to obtain it directly without completely solving the equation. For u and v ranging from 1 to 10 the results are given in Table 3.1.

It is seen that z_{sup} ranges from 1 to 2, as was to be expected (§ 3.4.2).

C) Special cases

Three particular cases deserve special mention.

*1. **First particular case***: $u = 0$ (only active operations on the loop path)

Then

$$z_{\text{sup}} = 2$$

The rate of growth (2^k) is as high as possible in a manner compatible with the assumption made on the cell operations which constitute the generating word (one cell generates at most two cells).

Example: system $DS_U = \dfrac{L\,L\,L^F}{a,b,c}$

It was found in Paragraph 3.4.1 that this system grows as 2^k (eq. 3-16). One has:

$$u = 0 \qquad v = 3 \qquad F(z) = \frac{1}{(z-1)^3}$$

$$a_0(z) = \frac{z}{z-1} \qquad a(z) = \frac{z(z-1)^2}{(z-1)^3 - 1}$$

$$b(z) = a(z)\frac{1}{z-1} = \frac{z(z-1)}{(z-1)^3 - 1} \qquad c(z) = a(z)\frac{1}{z-1}\,\frac{1}{z-1} = \frac{z}{(z-1)^3 - 1}$$

The poles of these three z-transforms are

$$\frac{1}{2} + \mathrm{i}\frac{\sqrt{3}}{2} \qquad \frac{1}{2} - \mathrm{i}\frac{\sqrt{3}}{2} \qquad 2 = z_{\text{sup}}$$

*2. **Second particular case:*** $v = 1$ (one active operation on the loop path)

Equation (3-31) then has the form

$$z^{u+1} - z^u - 1 = 0$$

It can be solved in a rigorous manner by means of Mellin's hypergeometric-function expansion (Belardinelli 1960, pp. 40, 56-57; Węgrzyn, Gille and Vidal 1985, Appendix B).

The denominator of $a(z)$ (eq. 3-29) and usually of some other $j(z)$ ($j = b,\ldots,n$) is

$$z^{u+1} - z^u - 1$$

which is characteristic of generalized *Fibonacci sequences* (if $u = 1$, of ordinary Fibonacci sequences): see examples below.

Table 3.1 Dominant root of equation (3-31) $z^u(z - 1)^v - 1 = 0$.

v =	1	2	3	4	5	6	7	8	9	10
$u = 1$	1.6180	1.7549	1.8192	1.8567	1.8813	1.8987	1.9116	1.9216	1.9296	1.9361
$u = 2$	1.4656	1.6180	1.7016	1.7549	1.7919	1.8192	1.8401	1.8567	1.8701	1.8813
$u = 3$	1.3803	1.5289	1.6180	1.6782	1.7218	1.7549	1.7809	1.8019	1.8192	1.8337
$u = 4$	1.3247	1.4656	1.5551	1.6180	1.6651	1.7016	1.7309	1.7549	1.7749	1.7919
$u = 5$	1.2852	1.4178	1.5056	1.5693	1.6180	1.6566	1.6880	1.7141	1.7361	1.7549
$u = 6$	1.2554	1.3803	1.4656	1.5289	1.5783	1.6180	1.6508	1.6782	1.7016	1.7218
$u = 7$	1.2321	1.3499	1.4324	1.4948	1.5442	1.5845	1.6180	1.6464	1.6708	1.6920
$u = 8$	1.2132	1.3247	1.4043	1.4656	1.5146	1.5551	1.5890	1.6180	1.6431	1.6651
$u = 9$	1.1975	1.3034	1.3803	1.4401	1.4886	1.5289	1.5631	1.5925	1.6180	1.6405
$u = 10$	1.1843	1.2852	1.3594	1.4178	1.4656	1.5056	1.5398	1.5693	1.5952	1.6180

Example I: system $DS_V = \dfrac{L\,T^F}{a,b}$ (eq. 3-17)

$$u = 1 \qquad v = 1 \qquad F(z) = \frac{1}{z(z-1)}$$

$$a_0(z) = \frac{z}{z-1} \qquad a(z) = \frac{z^2}{z^2 - z - 1}$$

$$b(z) = a(z)\frac{1}{z} = \frac{z}{z^2 - z - 1}$$

The z-transforms of *Fibonacci sequences* are recognized

$$a(k+2) = a(k+1) + a(k) \qquad a(0) = a(1) = 1$$

The growth occurs (see $u = v = 1$ in the Table) as 1.6180^k.

Example II: system $DS_W = \dfrac{L\,T\,T \ldots T\,T^F}{a,b,c,\ldots,t,u}$ (eq. 3-19)

$$u = n - 1 \qquad v = 1 \qquad F(z) = \frac{1}{z^{n-1}(z-1)} \qquad a_0(z) = \frac{z}{z-1}$$

$$a(z) = \frac{z^n}{z^n - z^{n-1} - 1} \qquad b(z) = \frac{a(z)}{z} \qquad \ldots \qquad u(z) = \frac{a(z)}{z^{n-2}}$$

The sequences $a(k)$, $b(k+1)$, ..., $u(k+n-2)$ are *generalized Fibonacci sequences*,

$$x(k+n) = x(k+n-1) + x(k) \qquad x(0) = x(1) = \ldots = x(n-1) = 1$$

The growth occurs as z_{sup}^k where z_{sup} appears in the first column of the Table. (All the poles of $a(z)$, $b(z)$, ..., $u(z)$ can be exactly evaluated.)

Example III: system $DS_R = \dfrac{L\,C\,S\,T^F}{a,b,c,d}$ (eq. 3-2)

$$u = 2 \qquad v = 1 \qquad F(z) = \frac{1}{z^2(z-1)} \qquad a_0(z) = \frac{z}{z-1}$$

$$a(z) = \frac{z^3}{z^3 - z^2 - 1} \qquad b(z) = a(z)\frac{1}{z} = \frac{z^2}{z^3 - z^2 - 1}$$

$$c(z) = a(z)\frac{1}{z}\frac{1}{z-1} = \frac{z^2}{(z-1)(z^3 - z^2 - 1)} \qquad d(z) = a(z)\frac{1}{z}\frac{1}{z} = \frac{z}{z^3 - z^2 - 1}$$

$a(k)$, $b(k)$ and $d(k)$ are *generalized Fibonacci sequences* (see Paragraph 3.4.1 *in fine*).

3. *Third particular case*: $v = 0$ (no active operation on the loop path)

In that exceptional case equation (3-31) reduces to

$$z^u - 1 = 0 \qquad \text{whence} \qquad z_{\text{sup}} = 1$$

The growth is *not exponential*; the stimulating effect of feedback on growth occurs *via* introducing an additional pole $z = 1$ into all the $D_j(z)$ and hence increasing the power of k at which the growth occurs.

Example: system $DS_X = \dfrac{C\,S\,T^F}{a,b,c}$

This system is obtained by introducing feedback into the developmental system (§ 2.1.1-A)

(2-1) $$DS_F = \frac{C\,S\,S}{a,b,c}$$

The sequence of the operations is

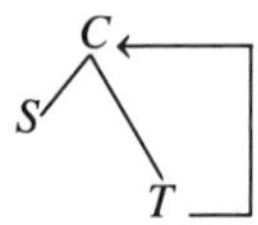

$a \rightarrow b(c)$

$b \rightarrow b$

$c \rightarrow a$

and the first developmental stages are

$$a \quad b(c) \quad b(a) \quad b(b(c)) \quad b(b(a)) \quad b(b(b(c))) \quad b(b(b(a)))$$

$$b(b(b(b(c)))) \quad b(b(b(b(a)))) \quad b(b(b(b(b(c))))) \quad b(b(b(b(b(a)))))$$

According to the general theory

$$u = 2 \qquad v = 0 \qquad F(z) = \frac{1}{z^2} \qquad a_0(z) = 1 \qquad a(z) = \frac{z^2}{z^2 - 1}$$

$$b(z) = a(z)\frac{1}{z-1} = \frac{z^2}{(z+1)(z-1)^2} \qquad c(z) = a(z)\frac{1}{z} = \frac{z}{z^2 - 1}$$

Taking the inverse transforms:

$$a(k) = \frac{1 + (-1)^k}{2} \qquad b(k) = \frac{2k + 1 + (-1)^{k-1}}{4} \qquad c(k) = \frac{1 + (-1)^{k-1}}{2}$$

The total number of cells at the k-th stage is

$$V(k) = \frac{k+2}{2} + \tfrac{1}{4}[1 + (-1)^{k-1}]$$

$$\{V(k)\} = \{1,2,2,3,3,4,4,5,5,6,6,7,7,8,8,9,9,\ldots\}$$

Whereas the size of system DS_F remained bounded (eq. 2-3), the introduction of feedback has caused system DS_X to grow indefinitely, the growth rate being not exponential in this exceptional case ($v = 0$), but polynomial (first power of k).

*

To summarize: the essential properties of synchronous developmental systems with feedback are the following, in contrast with systems without feedback.

1) Each cell has one antecedent.

2) The presence of feedback involves: (i) the presence of a loop in the generating word, which is of the circular type: (ii) the presence of a one in the first column of the evolution matrix, which is no longer upper-triangular.

3) After partial destruction, complete regeneration occurs if at least one cell associated to an operation lying inside the feedback loop has been preserved.

4) The system is extremely sensitive to mutation: changing one operation lying inside the feedback loop may completely alter the structure of the system.

5) The growth is exponential (except in one exceptional case). The rate depends on the number of active (L, R) and passive (B, C, T) operations in the feedback loop; it is 2^k if the latter comprises only active operations.

REFERENCES

G. BELARDINELLI (1960), *Fonctions hypergéométriques de plusieurs variables et Résolution analytique des équations algébriques générales*, Gauthier-Villars, Paris, 74 p.

B. DÉMIDOVITCH and I. MARON (1973), *Éléments de calcul numérique*, Mir, Moscow, 678 p.

F.R. GANTMACHER (1959), *The Theory of Matrices*, vol. 2, Chelsea, New York, 276 p.

J.C. GILLE, P. VIDAL and S. WĘGRZYN (1985), On some models for developmental systems. Part V: cellular composition of certain systems with feedback, transform approach, *International Journal of Systems Science*, vol. 16, pp. 1061-1072.

J.C. GILLE, P. VIDAL and S. WĘGRZYN (1987), On some models for developmental systems. Part VIII: systems with and without feedback, structural properties, *International Journal of Systems Science*, vol. 18, pp. 1195-1216.

J.C. GILLE, S. WĘGRZYN and P. VIDAL (1981), Chaînes créatrices bouclées, *Bulletin of the Polish Academy of Sciences (Technical Sciences)*, vol. 29, pp. 325-331.

J.C. GILLE, S. WĘGRZYN and P. VIDAL (1982), Développement des systèmes évolutifs à mot créateur bouclé, *Bulletin of the Polish Academy of Sciences* vol. 30, pp. 605-608.

J.C. GILLE, S. WĘGRZYN and P. VIDAL (1985), Grammaires des mots créateurs et propriétés des systèmes évolutifs correspondants étudiés par la transformation de Carson discrète, *Zentralblatt für Mathematik*, vol. 358, p. 350.

J.C. GILLE, S. WĘGRZYN and P. VIDAL (1986), Le principe de la régénération dans la théorie des systèmes évolutifs. Extrapolations biologiques, *Bulletin of the Polish Academy of Sciences (Technical Sciences)*, vol. 34, pp. 91-101.

G.T. HERMAN and G. ROZENBERG with a contribution by A. LINDENMAYER (1975), *Developmental Systems and Languages*, North-Holland Publishing Co., Amsterdam and American Elsevier, New York, 363 p.

R. JEAN (1978), *Phytomathématique*, Presses de l'Université du Québec, Montréal, 271 p.

J. LEGRAS (1963), *Précis d'analyse numérique*, Dunod, Paris, 256 p.

H.B. LÜCK and J. LÜCK (1976), Cellular number and cellular size in filamentous organisms in relation to ancestrally and positionally dependent generation times *in* A. LINDENMAYER and G. ROZENBERG (eds.), *Automata. Languages. Development*, North-Holland Publishing Co., Amsterdam and American Elsevier, New York, pp. 109-124.

H. MARCUS and H. MINC (1964), *A Survey of Matrix Theory and Matrix Inequalities*, Allyn and Bacon, Boston, 180 p.

A. PAZ and A. SALOMAA (1973), Integral sequential word functions and growth equivalence of Lindenmayer systems, *Information and Control*, vol. 23, pp. 313-343.

A. SALOMAA (1973), On exponential growth in Lindenmayer systems, *Indagationes mathematicae*, vol. 35, pp. 23-30.

A.L. SZILARD (1971), *Growth functions of Lindenmayer systems*, Technical Report 4, Dept. of Computer Sciences, University of Western Ontario, London (Ontario), 44 p.

P.M. VITÁNYI (1973), Structure of growth in Lindenmayer systems, *Indagationes mathematicae*, vol. 35, pp. 247-253.

S. WĘGRZYN, J.C. GILLE and P. VIDAL (1982), On some models for developmental systems. Part I: organism structure on the basis of elementary transformations, *International Journal of Systems Science*, vol. 13, pp. 359-378.

S. WĘGRZYN, J.C. GILLE and P. VIDAL (1984), Sur les propriétés des systèmes évolutifs engendrés par un mot créateur bouclé, *Revue française d'automatique, d'informatique et de recherche opérationnelle (Automatique)*, vol. 18, pp. 79-94.

S. WĘGRZYN, J.C. GILLE and P. VIDAL (1985), On some models for developmental systems. Part IV: cellular composition of certain systems with feedback, numerical and matrix approach, *International Journal of Systems Science*, vol. 16, pp. 1045-1060.

S. WĘGRZYN, J.C. GILLE and P. VIDAL (1986), On some models for developmental systems. Part VII: on some types of patterns included in a feedback loop, *International Journal of Systems Science*, vol. 17, pp. 569-579.

S. WĘGRZYN, P. VIDAL and J.C. GILLE (1984), Sur la relation entre le système évolutif $E = XS$ et le système bouclé associé $F = XT$, *Podstawy sterowania*, vol. 14, pp. 279-286.

S. WĘGRZYN, P. VIDAL and J.C. GILLE (1986), Mutations et théorie des systèmes évolutifs, *Bulletin of the Polish Academy of Sciences (Technical Sciences)*, vol. 34, pp. 121-124.

Chapter 4

DEVELOPMENTAL SYSTEMS WITH OPERATING SYSTEM

4.1 GENERAL

The model for developmental systems commented on in Chapters 2 and 3 is based on the idea that the development of a system is the result of elementary operations acting on its elements. It is in agreement with the following assumptions of cellular biology: (1) organisms are composed of cells, (2) each cell is derived from a preexisting cell and (3) a cell can give rise to at most two cells. This is the reason why it accounts for many facts concerning living organisms: quantitative growth, internal hierarchy, regeneration, cloning, mutation.

However the said model implicitly assumes that cells are independent of one another, which is not the case for living organisms. As a consequence it is subjected to two limiting hypotheses:

a) *synchronicity*, i.e. all A_i (a_i) operations are performed simultaneously at each developmental step;

b) *context-free* development, i.e. the operations grouped in the generating word are executed after one another independently of external circumstances.

In order better to account for the complex facts of living organisms a more elaborate model will now be proposed which is not limited by the above two assumptions. It was suggested by our present knowledge of the DNA structure, in particular by Jacob and Monod's theory of the regulation of protein synthesis by DNA in *E. Coli* (e.g. Suzuki *et al.*, 1981, pp. 585-608; Watson, 1976, pp. 379-410): the action of the structural (or developmental) genes which are responsible for protein synthesis is controlled by other genes that initiate or block the synthesis (Fig. 4-1) depending on the context (e.g. on temperature, on the chemical cell composition). These latter genes are located on DNA in a control area before the structural genes (in the sense that the DNA program is read from left to right).

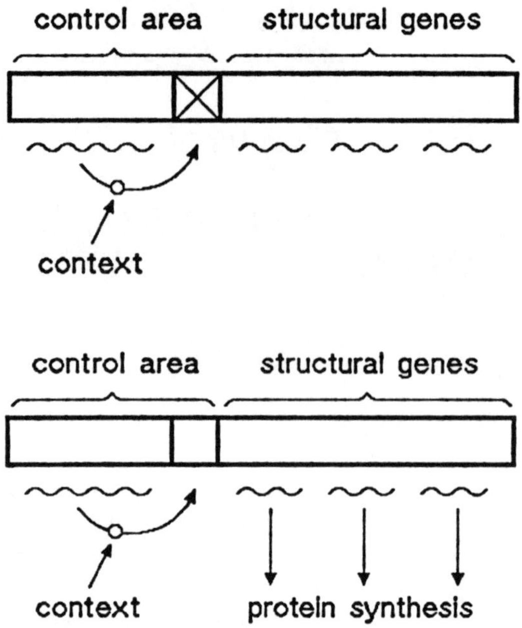

Fig. 4-1. Control of protein synthesis by DNA. Depending on context, synthesis is blocked (above) or takes place (below).

These facts have suggested to us a generalization of the initial model. It consists of having each A_i elementary operation preceded by a control operation O_i, thus resulting in a generalized elementary operation denoted

$$O_i A_i \tag{4-1}$$

The A_i (structural part) is responsible for the transformations undergone by the a_i cell (see Paragraph 1.3.2); the O_i controls when and why (depending on time and on internal and external context) the A_i operation is initiated or blocked.

Examples were quoted in Paragraph 1.2.5.

If an L operation acting on an a cell is performed only six times and then the a cell remains unaltered, we note

$$O_a L(a) = \begin{cases} ba & \text{if} \quad n_a \leqslant 6 \\ b & \text{if} \quad n_a > 6 \end{cases} \tag{1-15}$$

where n_a is the number of times the L operation has been performed. For brevity's sake we shall simply write

$$6L$$

If a b cell undergoes the S operation ten times and then disappears

(1-16) $$O_b S(b) = \begin{cases} b & \text{if} \quad n_b \leq 10 \\ - & \text{if} \quad n_b > 0 \end{cases}$$

where n_b is the number of times the S operation has been performed. More briefly we write

$$10S$$

A generating word made up of generalized operations will be called a *generalized generating word*, e.g.

$$GW = O_a L\, O_b S \qquad \text{or} \qquad 6L\ 10S \tag{4-2a}$$

$$GW = A_1\, O_2\, A_2\, A_3\, O_4\, A_4 \tag{4-3a}$$

To make the reading easier, parentheses will be introduced which isolate the two symbols of each generalized operation, i.e., the above generating words will be written

$$GW = (O_a L)\,(O_b S) \qquad \text{or} \qquad (6L)\,(10S) \tag{4-2b}$$

$$GW = A_1(O_2 A_2) A_3 (O_4 A_4) \tag{4-3b}$$

Note that the said parentheses introduce no additional information. Like in computers, expressions of the (4-2a, 4-3a) type, consisting of a sequence of operation symbols and argument symbols without parentheses ("Polish notation"), should be read directly from left to right.

To summarize: introducing control operations O_i into a simple generating word

(1-18) $$GW = \underset{i=1}{\overset{n}{\mathrm{SEQ}}}\, A_i\,(a_i)$$

results in a generalized generating word

(1-27) $$GW = \underset{i=1}{\overset{n}{\mathrm{SEQ}}}\, O_i A_i\,(a_i) \tag{4-4}$$

If the generating word is looked upon as the program of a computer, the O_i constitute an *operating system*, as in many multiprogrammed processes.

4.2 SENSITIVITY TO EXTERNAL CONTEXT

Context-bound development of a system can be accounted for by a *real-time operating system*.

Consider for example the developmental system DS_G (eq. 2-4) which was found in Paragraph 2.2.1-B to have the shape of a tree with indefinitely growing branches. Now suppose that the system environment is characterized by the succession of day and night and some operations present in the generating word are performed only during day time (subscript D) whereas others are performed only at night (subscript N):

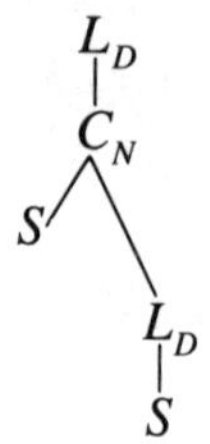

$$\begin{array}{ll} a \rightarrow ba & \text{Day} \\ b \rightarrow c(d) & \text{Night} \\ c \rightarrow c & \\ d \rightarrow ed & \text{Day} \\ e \rightarrow e & \end{array} \qquad (4\text{-}5)$$

The generalized generating word is

$$GW = (O_aL)\,(O_bC)\,S\,(O_dL)\,S$$

and the developmental system is

$$DS_y = \frac{(O_aL)\,(O_bC)\,S\,(O_dL)\,S}{\text{a,b,c,d,e}}$$

where

$$O_aL(a) = \begin{cases} ba & \text{Day} \\ a & \text{Night} \end{cases}$$

$$O_bC(b) = \begin{cases} c(d) & \text{Night} \\ c & \text{Day} \end{cases}$$

$$O_dL(d) = \begin{cases} ed & \text{Day} \\ e & \text{Night} \end{cases}$$

In day time the development is:

$$a \qquad ba \qquad bba \qquad bbba \qquad bbbba \qquad bbbbba \qquad \ldots$$

(to be specific it is assumed that five b have been generated). Then at night

$$c(d)c(d)c(d)c(d)c(d)a$$

is obtained. Hence the next day

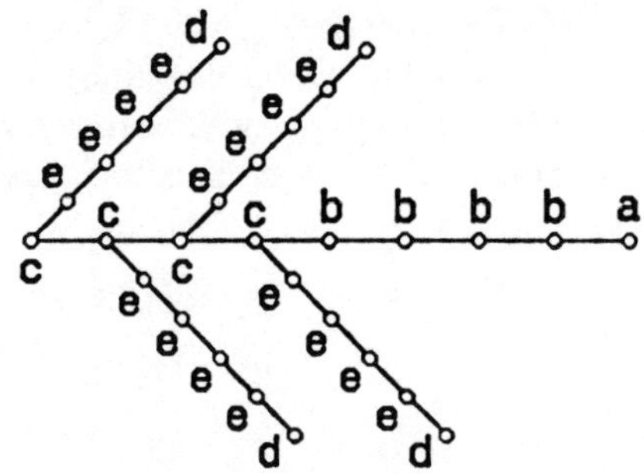

Fig. 4-2. Modifications of the development of system DS_G when a real-timc operating system causes the L operations to be active during day time and the C operation to be active at night.

$$\left.\begin{array}{c} c(ed)c(ed)c(ed)c(ed)c(ed)ba \\ c(eed)c(eed)c(eed)c(eed)c(eed)bba \\ c(eeed)c(eeeed)c(eeed)c(eeed)bbba \\ c(eeeed)c(eeeed)c(eeeed)c(eeeed)bbbba \end{array}\right\} \quad (4\text{-}6)$$

It is observed (Fig. 4-2) that the presence of the operating system has modified the shape of the developmental system: the respective dimensions of the "organs" (stem, branches) have been modified, but the fundamental structure has not been altered.

4.3 SENSITIVITY TO INTERNAL CONTEXT

Examples will now be given of operating systems which control the execution of some of the A_i operations (*i*) either according to the number of times the (or another) operation has been performed (*ii*) or according to the number of cells of a given category which are present in the system or have been generated by a certain operation.

4.3.1 Example I: limitation of cell number

Consider again the developmental system

$$DS_G = \frac{L\,C_{\pm 45}\,L\,S\,S}{a,b,c,d,e} \quad (2\text{-}4)$$

(tree with indefinitely growing branches). Now suppose that an operating system modifies the execution of the two L operations in such a manner that the number of cells generated by each of the L operations is limited to four, i.e. the L operations become stagnations S after they have been active four times. In other words

$$O_a L(a) = \begin{cases} ba & \text{if} \quad n_a \leqslant 4 \\ a & \text{if} \quad n_a > 4 \end{cases}$$

$$O_d L(d) = \begin{cases} ed & \text{if} \quad n_d \leqslant 4 \\ e & \text{if} \quad n_d > 4 \end{cases}$$

The system thus modified is

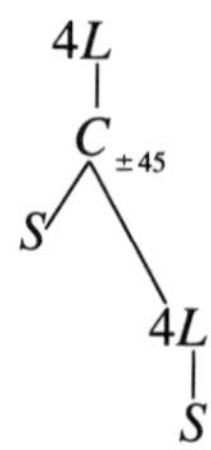

$$\begin{array}{ll} a \rightarrow ba & \text{limited to 4 operations} \\ b \rightarrow c(d) & \\ c \rightarrow c & \\ d \rightarrow ed & \text{limited to 4 operations} \\ e \rightarrow e & \end{array} \qquad (4\text{-}7)$$

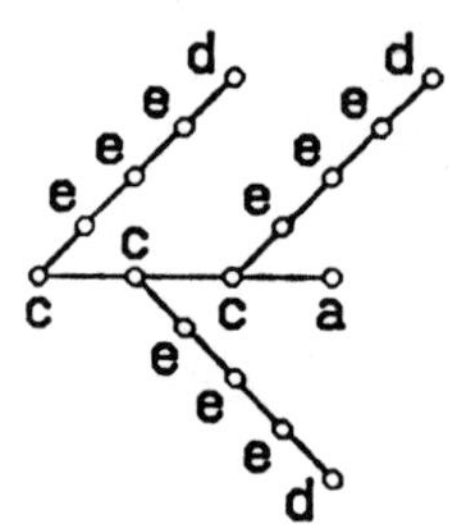

Fig. 4-3. Limitation of branch growth.

The first developmental stages are

$$\left.\begin{array}{l} DS_Z(0) = a \qquad DS_Z(1) = ba \qquad DS_Z(2) = c(d)ba \\ DS_Z(3) = c(ed)c(d)ba \qquad DS_Z(4) = c(eed)c(ed)c(d)ba \\ DS_Z(5) = c(eeed)c(eed)c(ed)a \\ DS_Z(6) = c(eeed)c(eeed)c(eed)a \\ DS_Z(7) = c(eeed)c(eeed)c(eeed)a \\ DS_Z(8) = DS_Z(9) = \ldots = DS_Z(7) \end{array}\right\} \qquad (4\text{-}8)$$

Figure 4-3 shows the final state of the system ($k > 7$): it is observed that the number of the branches is limited and their length remains equal to 4 cells.

4.3.2 Other examples: semi-synchronous systems

Systems will now be considered in which the $A_i(a_i)$ operations are performed at instants which differ from one another but are all instants at which a new developmental stage is initiated. Such a strategy characterizes what may be called a *semi-synchronous* system.

A) Slowing down of branch growth

Considering the system DS_H (tree with indefinitely growing branches)

$$DS_H = \frac{L\,C_{\pm 45}\,S\,L\,S}{a,b,c,d,e} \tag{2-7}$$

again, it is possible to slow down the growth of the branches without altering the growth of the stem by prescribing that the second L operation ($d \rightarrow ed$) be performed only after the first L operation ($a \rightarrow ba$) and the C operation [$b \rightarrow c(d)$] have been performed four times. The generating word will be written

$$L\,C\,S\,(OL)\,S \tag{4-9a}$$

and the developmental system DS_{AA}

$$DS_{AA} = \frac{L\,C\,S\,(O_dL)\,S}{a,b,c,d,e}$$

where O_dL should be understood as follows:

$$O_dL(d) = \begin{cases} d & n_L < 4 \\ ed & n_L \geqslant 4 \end{cases}$$

where n_L is the number of times the first L operation has been performed.

The operation sequence is

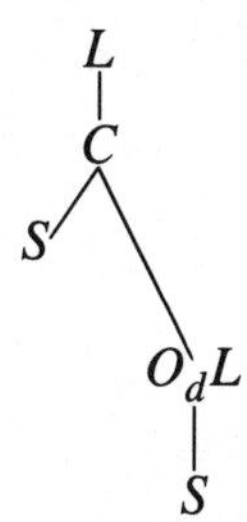

$$\begin{aligned} a &\rightarrow ba \\ b &\rightarrow c(d) \\ c &\rightarrow c \\ d &\rightarrow \begin{cases} d \\ ed \end{cases} \\ e &\rightarrow e \end{aligned} \tag{4-9b}$$

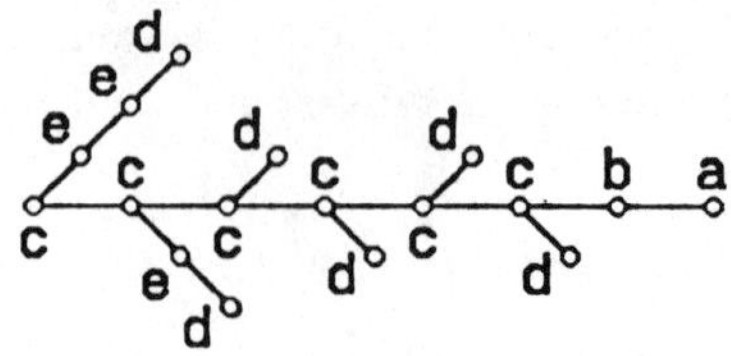

Fig. 4-4. Slowing down of branch growth.

The first developmental stages are

$$DS_{AA}(0) = a \qquad DS_{AA}(1) = ba \qquad DS_{AA}(2) = c(d)ba$$

$$DS_{AA}(3) = c(d)c(d)ba \qquad DS_{AA}(4) = c(d)c(d)c(d)ba$$

$$DS_{AA}(5) = c(d)c(d)c(d)c(d)ba$$

At $k = 6$ the condition for O_dL to be an L operation has been fulfilled so far as the first d is concerned, whence

$$DS_{AA}(6) = c(ed)c(d)c(d)c(d)c(d)ba$$

At $k = 7$ this applies also to the second d:

$$DS_{AA}(7) = c(eed)c(ed)c(d)c(d)c(d)c(d)ba \tag{4-10}$$

(see Figure 4-4), etc.

B) Slowing down of branch generation

Suppose now that the $C_{\pm 45}$ $[b \to c(d)]$ operation is executed only after the first L $(a \to ba)$ operation has been performed four times:

$$DS_{BB} = \frac{L\,(O_b\,C)\,S\,L\,S}{a,b,c,d,e} \tag{4-11}$$

where

$$O_bC(b) = \begin{cases} b & \text{if} \quad n_L < 4 \\ c(d) & \text{if} \quad n_L \geq 4 \end{cases}$$

n_L = number of times the first L operation has been performed.

The operation sequence is

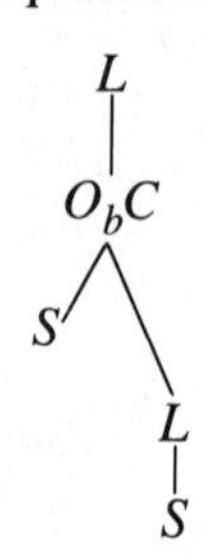

$$a \to ba$$

$$b \to \begin{cases} b \\ c(d) \end{cases}$$

$$c \to c \tag{4-12}$$

$$d \to ed$$

$$e \to e$$

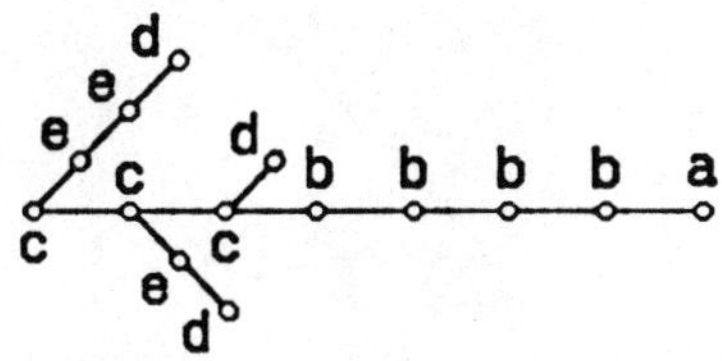

Fig. 4-5. Slowing down of branch generation.

During the first four stages only the stem grows ($O_bC = S$):

$$DS_{BB}(0) = a \qquad DS_{BB}(1) = ba \qquad DS_{BB}(2) = bba$$

$$DS_{BB}(3) = bbba \qquad DS_{BB}(4) = bbbba$$

Then O_bC becomes a C operation, whence

$$DS_{BB}(5) = c(d)bbbba \qquad DS_{BB}(6) = c(ed)c(d)bbbba$$

$$DS_{BB}(7) = c(eed)c(ed)c(d)bbbba$$

The $k = 7$ pattern is shown in Figure 4-5.

C) SLOWING DOWN OF STEM GROWTH

Suppose finally that the first L operation is executed (1) for $k = 1, 2, 3$ and (2) later only after the second L operation ($d \to ed$) has been performed four times:

$$DS_{CC} = \frac{(O_aL)\ C\ S\ L\ S}{a,b,c,d,e}$$

where

$$O_aL(a) = \begin{cases} b(a) & k = 1,\ 2,\ 3 \\ a & k > 3 \quad n_L < 4 \\ b(a) & n_L \geq 4 \end{cases}$$

The operation sequence is

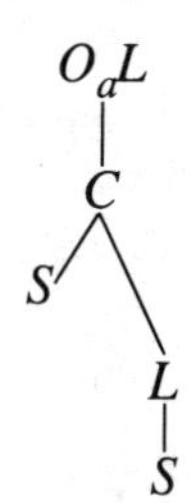

$$a \to \begin{cases} b(a) \\ a \end{cases}$$

$$b \to c(d)$$

$$c \to c$$

$$d \to ed$$

$$e \to e$$

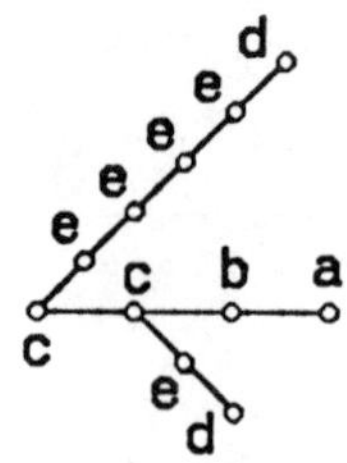

Fig. 4-6. Slowing down of stem growth.

The first six stages of development are (see Figure 4-6)

$$\left.\begin{array}{l} DS_{CC}(0) = \mathrm{a} \qquad DS_{CC}(1) = \mathrm{ba} \qquad DS_{CC}(2) = c(d)ba \\ DS_{CC}(3) = c(ed)c(d)ba \qquad DS_{CC}(4) = c(eed)c(ed)a \\ DS_{CC}(5) = c(eeed)c(eed)a \\ DS_{CC}(6) = c(eeeed)c(eeed)a \end{array}\right\} \qquad (4\text{-}14)$$

D) Communication between cells

There exists inside a living organism a solidarity between the cells. The cells are interrelated between one another from the triple viewpoint of energy, of substance and of information, the activity of any cell being influenced by the state of other cells. Such exchanges are performed through the cellular membrane (Popot, 1987) and by molecules which circulate from one cell to another (Berridge, 1985; Snyder, 1985).

In our six-operation model such a type of coupling can be realized by an operating system which prescribes that the A_i operation on the a_i cell should be performed or not depending on a message originating from the a_k cell $(k \neq i)$:

$$O(a_k)_i A_i(a_i) = \begin{cases} a_i & \text{in the absence of any message from } a_k \\ A_i(a_i) & \text{in the presence of a message from } a_k \end{cases}$$

This will be illustrated by the simple example of a developmental system consisting of two subsystems each of which has a generating subword of the *LCSLS* (eq. 2-7) type: one subsystem (the "stem") is directed upward, the other (the "root") is directed downward.

The initial cell will be denoted α. The cells which constitute the system will bear index 1 when they belong to the stem $(a_1,\ldots,e_1)$ and index 2 when they belong to the root $(a_2,\ldots,e_2)$. Synchronization between the two subsystems will be obtained by assuming that the initial cell a_1 of the stem generates another cell $[D(a_1) \rightarrow b_2 a_1]$ only once an e_2 cell has appeared in the root. In other words the development of the stem is postponed [stagnation $S(a_1) \rightarrow a_1$] so long as the root has not developed enough.

The system is thus defined as

$$DS_{\mathrm{DD}} = \frac{B[O(e_2)L]\,L\,C\,C\,S\,L\,S\,L\,S\,S}{\alpha,a_1,b_1,c_1,d_1,e_1,a_2,b_2,c_2,d_2,e_2} \tag{4-16}$$

where

$$O(e_2)L(a_1) = \begin{cases} a_2 a_1 & \text{if at least one } e_2 \text{ element is present} \\ a_1 & \text{if no } e_2 \text{ element is present} \end{cases}$$

The sequence of the elementary operations is the following:

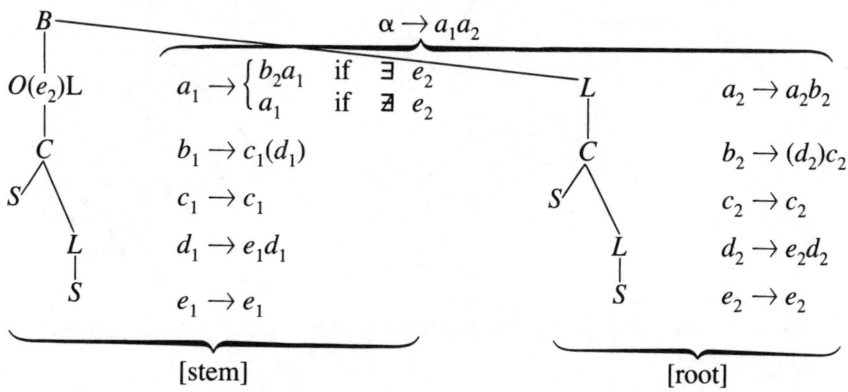

The following developmental stages, starting from one initial cell, are obtained. (Recall that the "root" grows downward and the stem grows upward; the instructions are to be read from right to left for the former and from left to right for the latter.) See Figure 4-7.

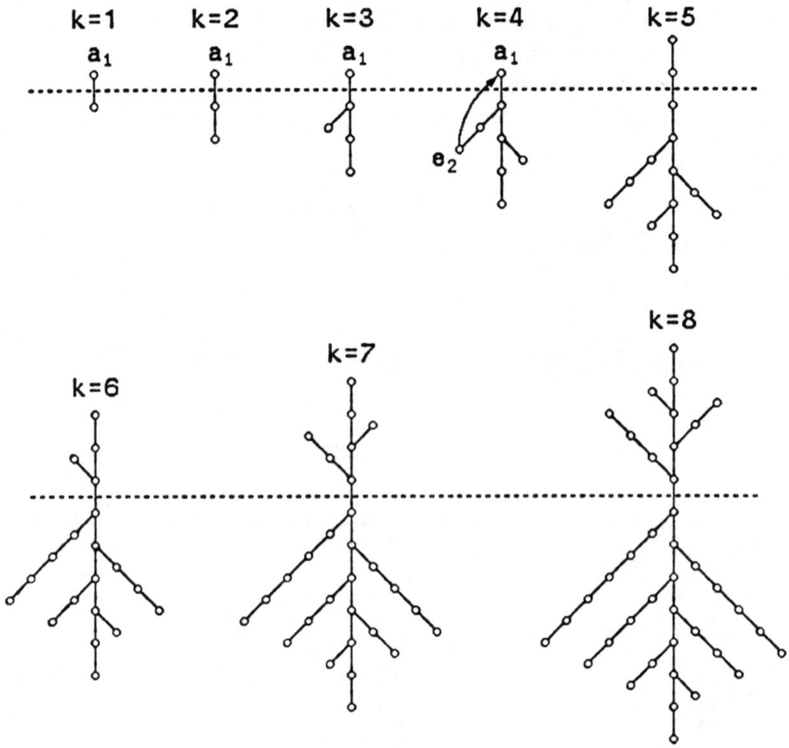

Fig. 4-7. The stem starts developing only after one e_2 cell has appeared in the root.

$$
\begin{array}{rll}
k = 0 & \alpha & \\
k = 1 & a_2 & a_1 \\
k = 2 & a_2b_2 & a_1 \\
k = 3 & a_2b_2(d_2)c_2 & a_1 \\
k = 4 & a_2b_2(d_2)c_2(d_2\mathbf{e}_2)c_2 & a_1 \\
k = 5 & a_2b_2(d_2)c_2(d_2e_2)c_2(d_2e_2e_2)c_2 & b_1a_1 \\
k = 6 & a_2b_2(d_2)c_2(d_2e_2)c_2(d_2e_2e_2)c_2(d_2e_2e_2e_2)c_2 & c_1(d_1)b_1a_1 \\
k = 7 & \underbrace{a_2b_2(d_2)c_2(d_2e_2)c_2(d_2e_2e_2)c_2(d_2e_2e_2e_2)c_2 \; (d_2e_2e_2e_2e_2)c_2}_{\text{root}} & \underbrace{c_1(e_1d_1)c(d_1)b_1a_1}_{\text{stem}}
\end{array}
$$

It is observed that the stem development is blocked during the first stages: since no e_2 cell yet exists in the root, the $0(e_2)L$ operation acting on the a_1 cell is a stagnation (see equation 4-12). The situation changes at the $k = 4$ developmental stage: the root then has grown enough, more precisely, an e_2 cell has been generated; as a consequence the $0(e_2)L$ operation becomes a linear generation (eq. 4-12), i.e. the stem starts developing in its turn.

In all the above examples it should be noted that the presence of the operating system modifies the respective proportions of the system components but does not alter its structure.

4.4 DWINDLING AND REGROWTH

The developmental systems considered till now (except DS_E of § 1.2.5) grow monotonically and indefinitely. It will now be shown that a model with operating system enables one to account for the property of living organisms to decay after they have reached their "adult" size and to reproduce in the form of a new or of many new system(s).

4.4.1 Ephemeral developmental systems

In a developmental system the generating word of which has no operating system the A_i operation is repeated so long as at least one a_i cell is present. Therefore no decay ever appears. In fact: (1) if at least one L or R operation is present in the generating word the growth is indefinite, and (2) if there is none, the growth stops and the system indefinitely keeps the maximum size it has reached, as a consequence of the S operations.

This suggests that developmental systems which decay after they have reached their maximum size may be obtained if the generating word contains an operating system which prescribes that the *generative operations L, R* and the *stagnation operations S* be performed only a *limited number of times*. The said operations will be noted *mL*, *mR* and *mS* respectively. They are defined as follows.

1/ The *mL* or *mR* operation is defined by (see eq. 1-15)

$$a_i \rightarrow \begin{cases} a_j a_i \quad \text{or} \quad a_j(a_i) & n_{a_i} \leq m \\ a_j & n_{a_i} > m \end{cases} \tag{4-17}$$

where n_{a_i} is the number of times the operation has been performed on the a_i cell and on the descendants thereof.

2/ The *mS* operation is defined by (see eq. 1-16)

$$a_i \rightarrow \begin{cases} a_i & n_{a_i} \leq m \\ - & n_{a_i} > m \end{cases} \tag{4-18}$$

where n_{a_i} is the number of developmental stages during which a_i has been present in the system and the dash means that the a_i cell disappears ("dies").

If the generating word is linear, it can be easily seen that such a system *does not grow indefinitely*: after a certain number of developmental steps its size decreases and finally the system dwindles away.

This will be illustrated by the following developmental system

$$X = \frac{(4L)\, C\, (14S)\, 3L\, (9S)}{a,b,c,d,e} \tag{4-19}$$

(adapted from DS_H, eq. 2-7), the development of which is controlled by the following operations:

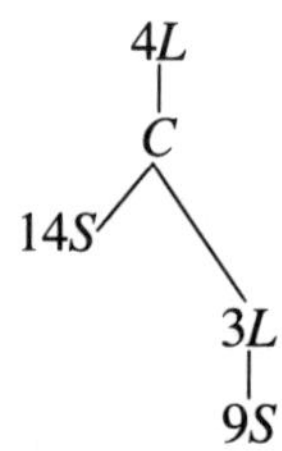

$$\begin{aligned} &a \rightarrow 4 \quad \text{times } ba, \text{ then } b \\ &b \rightarrow c(d) \\ &c \rightarrow 14 \quad \text{times } c, \text{ then } - \\ &d \rightarrow 3 \quad \text{times } ed, \text{ then } d \\ &e \rightarrow 9 \quad \text{times } e, \text{ then } - \end{aligned} \tag{4-20}$$

The successive developmental stages starting from one initial a cell can be computed step by step on the basis of the above relations (4-20). It is found that

$$X(0) = a \qquad X(1) = ba \qquad X(2) = c(d)ba$$

$$X(3) = c(ed)c(d)ba \qquad X(4) = c(eed)c(ed)c(d)ba$$

At the next step, the $4L$ operation, which has acted four times on the a cell, acts as $a \rightarrow b$. Thus

$$X(5) = c(eeed)c(eed)c(ed)c(d)b$$

At the next step the $3L$ operation, which has acted three times on the d cell, acts as $d \rightarrow e$. Thus

$$X(6) = c(eeee)c(eeed)c(eed)c(ed)c(d)$$

The system still grows, as a consequence of the $3L$ operation:

$$X(9) = c(eeee)c(eeee)c(eeee)c(eeee)c(eeed)$$

Then

$$X(10) = \mathrm{X}(11) = \mathrm{X}(12) = \mathrm{c(eeee)c(eeee)c(eeee)c(eeee)c(eeee)}$$

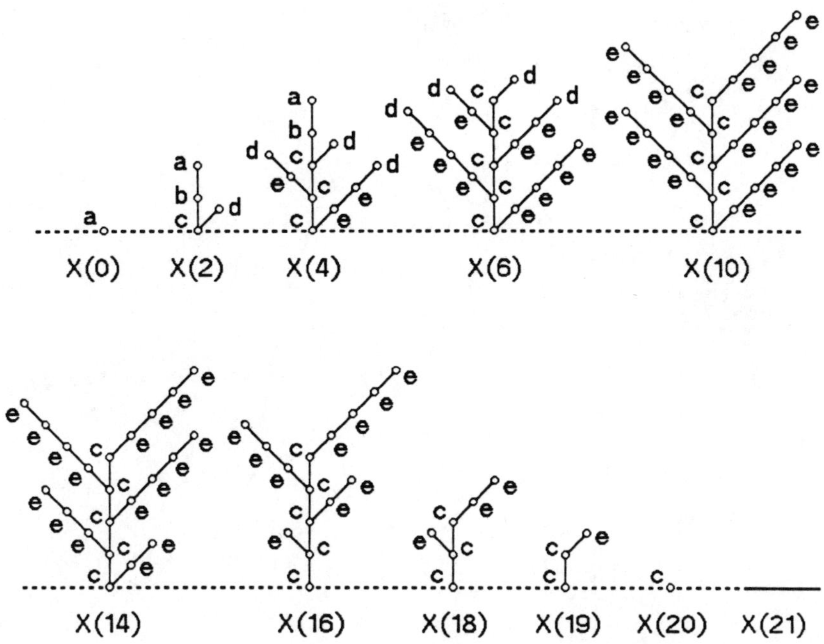

Fig. 4-8. Growth and decay of a developmental system.

After $X(12)$ the e cells gradually vanish away, as a consequence of the $9S$ operation when $n_e > 9$:

$$X(13) = c(eee)c(eeee)c(eeee)c(eeee)c(eeee)$$

$$X(14) = c(ee)c(eee)c(eeee)c(eeee)c(eeee)$$

$$X(15) = c(e)c(ee)c(eee)c(eeee)c(eeee)$$

Then the c cells vanish in their turn ($14S$ operation):

$$X(18) = cc(e)c(ee) \qquad X(19) = cc(e) \qquad X(20) = c$$

and for $k \geqslant 21$ the system disappears.

The patterns at some typical developmental stages are shown in Figure 4-8. The manner in which the system size (i.e., the number of cells) varies as a function of k is shown in Figure 4-9.

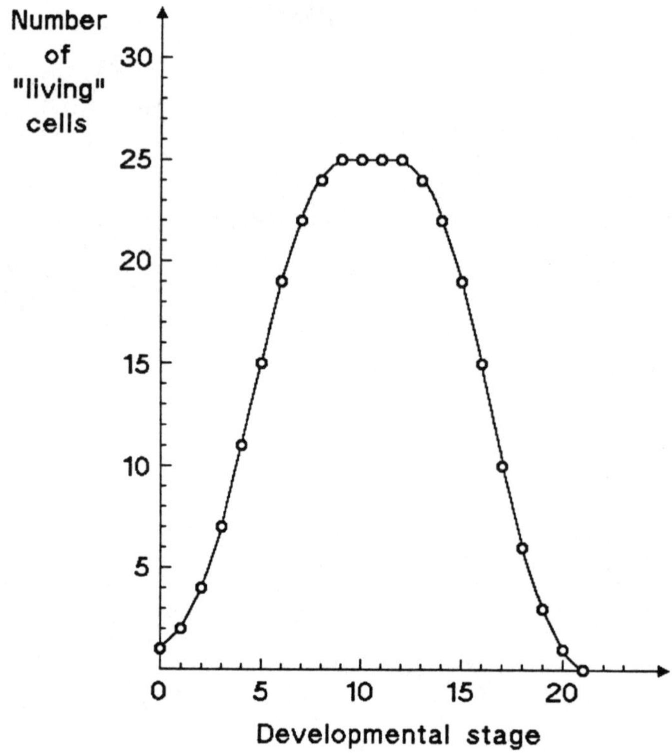

Fig. 4-9. Number of "living" cells of developmental system (4-19) (see Figure 4-8).

4.4.2 "Reproduction" of a developmental system

Systems with a *circular generating word* will now be analyzed.

A) Simple reproduction

Consider a developmental system differing from the X system (eq. 4-19) by the presence of an additional first operation B and of an additional last operation T with *feedback* from the latter towards the former. Furthermore, suppose that the T operation is performed only after the 25th developmental stage; it will therefore be denoted $D25T$, where D stands for "delay":

$$D25T(a_i) = \begin{cases} a_i & n_{a_i} < 25 \\ a_j & n_{a_i} = 25 \end{cases} \tag{4-21}$$

The developmental system thus constructed is

$$Y = \frac{B\,X\,(D25T)^F}{\alpha,a,b,c,d,\beta} = \frac{B\,(4L)\,C\,(14S)\,4L\,(9S)\,(D25T)^F}{\alpha,a,b,c,d,\beta} \tag{4-22}$$

which expresses the following operations:

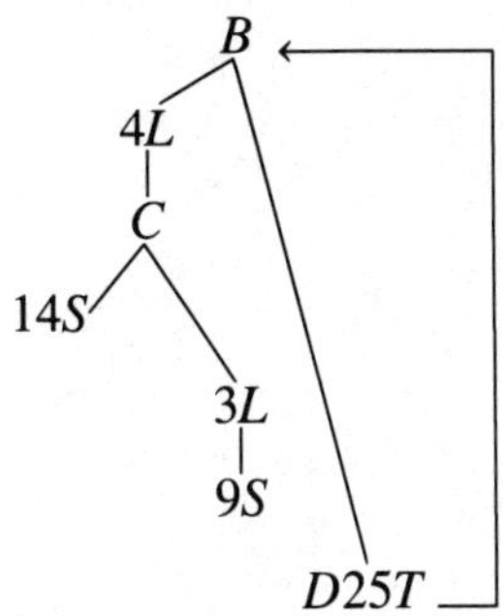

$$\begin{aligned}
&\alpha \to a\beta\\
&a \to 4 \text{ times } ba, \text{ then } b\\
&b \to c(d)\\
&c \to 14 \text{ times } c, \text{ then } -\\
&d \to 3 \text{ times } ed, \text{ then } e\\
&e \to 9 \text{ times } e, \text{ then } -\\
&\beta \to \alpha \text{ with 25 stage delay}
\end{aligned} \tag{4-23}$$

From eqs. (4-23) the successive developmental stages starting from the initial condition $Y(0) = \alpha$ are easily computed (Fig. 4-10). The first stages are quite comparable to the development of the X system of the foregoing paragraph:

$$Y(0) = \alpha \qquad Y(1) = a\beta = x(0)\beta \qquad Y(2) = ba\beta = X(1)\beta$$

$$Y(3) = c(d)ba\beta = X(2)\beta \qquad Y(4) = c(ed)c(d)ba\beta = X(3)\beta$$

i.e.

$$Y(\mathrm{i}) = X(i-1)\beta \qquad 1 \leqslant i \leqslant 25$$

Thus

$$Y(20) = cc(e)\beta = X(19)\beta \qquad Y(21) = c\beta = X(20)\beta$$

$$Y(22) = Y(23) = Y(24) = Y(25) = \beta$$

Then the $D25T$ operation becomes active:

$$Y(26) = \alpha = Y(0)$$

From then on a *new developmental cycle* begins, identical to the first one:

$$Y(27) = e\beta = Y(1) \qquad Y(28) = ba\beta = Y(2)$$

in general

$$Y(i) = Y(i-26) \qquad 26 \leqslant i \leqslant 52$$

Overall, the development is *periodic* with a period equal to 26 steps. During each period the system grows, then dwindles down to one cell — from which "germ" the next cycle starts. In a certain sense, such a system can be termed "immortal".

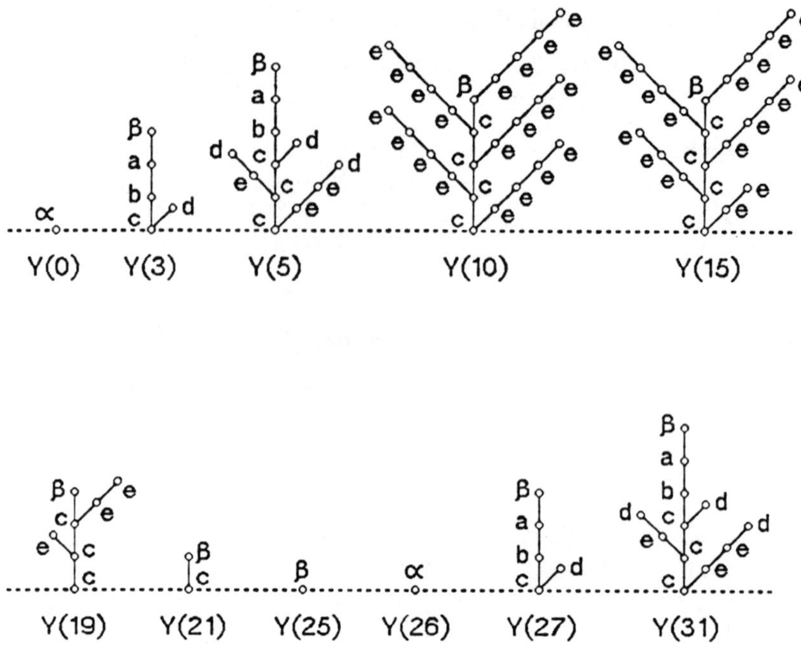

Fig. 4-10. Dwindling and regrowth of a developmental system.

B) MULTIPLE REPRODUCTION

Now suppose that feedback has been introduced into the generating word of the X developmental system (eq. 4-19) from a final $D25T$ operation, but the B operation which was present in the Y system (eq. 4-22) is no more the first operation:

$$Z = \frac{(4L)\ C\ (14S)\ B\ (4L)\ (9S)\ (D25T)^F}{a,b,c,\alpha,d,e,\beta} \tag{4-24}$$

The operation sequence is

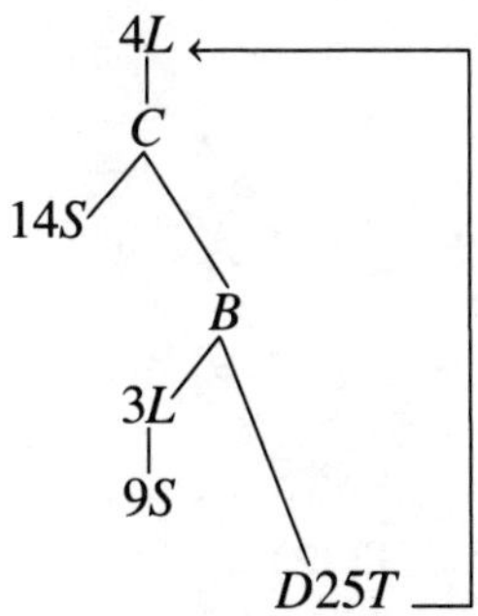

$$\begin{aligned}
&a \to 4 \quad \text{times } ba, \text{ then } b\\
&b \to c(\alpha)\\
&c \to 14 \quad \text{times } c, \text{ then } -\\
&\alpha \to d\beta\\
&d \to 3 \quad \text{times } ed, \text{ then } e\\
&e \to 9 \quad \text{times } e, \text{ then } -\\
&\beta \to a \text{ with a 25 stage delay}
\end{aligned} \tag{4-25}$$

The first developmental stages are (Fig. 4-11)

$$Z(0) = a \qquad Z(1) = ba \qquad Z(2) = c(\alpha)ba$$

$$Z(3) = c(d\beta)c(\alpha)ba \qquad Z(4) = c(ed\beta)c(d\beta)c(\alpha)ba$$

after which the $4L$ operation acts as $a \to b$:

$$Z(5) = c(eed\beta)c(ed\beta)c(d\beta)(c(\alpha)b$$

Then

$$Z(6) = c(eeed\beta)c(eed\beta)c(eed\beta)c(ed\beta)c(d\beta)$$

During the next four steps the $3L$ operation acts on certain d as $d \to e$:

$$Z(7) = c(eeee\beta)c(eeed\beta)c(eed\beta)c(ed\beta)c(d\beta)$$

$$\ldots$$

$$Z(11) = Z(12) = Z(13) = c(eeee\beta)c(eeee\beta)c(eeee\beta)c(eeee\beta)c(eeee\beta)$$

After $k = 13$ the e cells [operation $9S$] gradually disappear, as do the c cells after $k = 16$ [operation $14S$], so that the system size decreases:

$$Z(14) = c(eee\beta)c(eeee\beta)c(eeee\beta)c(eeee\beta)c(eeee\beta)$$

$$Z(16) = c(\beta)c(e\beta)c(ee\beta)c(eee\beta)c(eeee\beta)$$

$$Z(18) = (\beta)(\beta)(e\beta)c(ee\beta)c(eee\beta)$$

$$Z(20) = (\beta)(\beta)(\beta)(\beta)(\mathrm{e}\beta)$$

until the system is reduced to five separate cells:

$$Z(21) = \ldots = Z(27) = (\beta)(\beta)(\beta)(\beta)(\beta)$$

At this stage the $D25T$ operation becomes active:

$$Z(28) = (a)(a)(a)(a)(a)$$

Each of the *five* separate a cells ("germs") then gives rise to a *new cycle*, identical to the previous one.

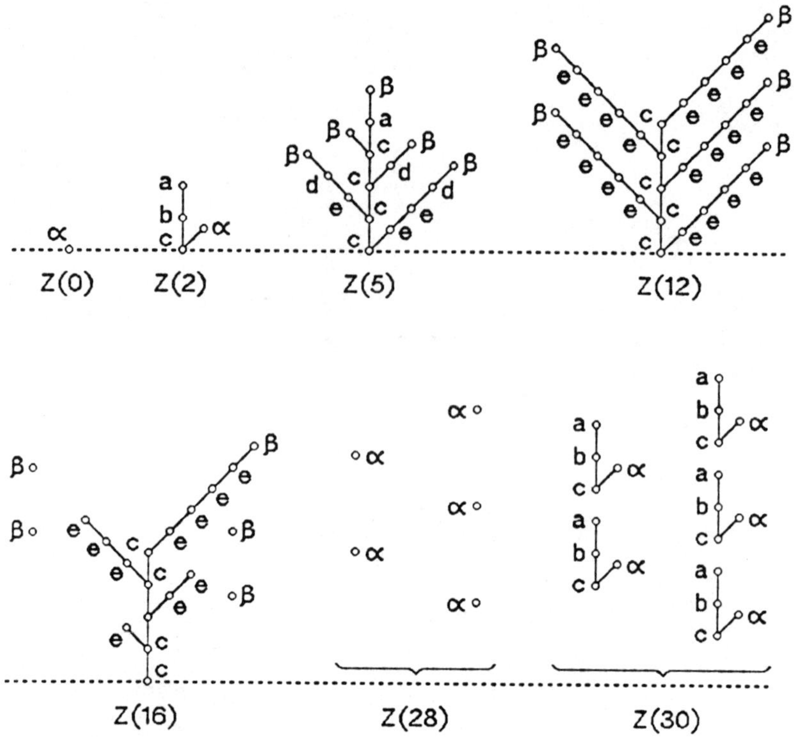

Fig. 4-11. Dwindling and reproduction of a developmental system.

Altogether the development is *periodic*, with a period of 28 steps and *multiplication* by 5 at each cycle.

4.5 MULTILEVEL DEVELOPMENT

4.5.1 General

It is possible to construct developmental systems the development of which occurs in several successive phases, in the sense that one part develops first and then remains "dormant" while other parts develop in their turn (Węgrzyn, Vidal and Gille, 1990).

Consider for example the system

$$DS_{EE} = \frac{(0_a \mathrm{L})\, S\, B\, (0_d R)\, S\, (0_e R)\, S}{a,b,c,d,e,f,g,h} \tag{4-26}$$

where

$$0_aL(a) = \begin{cases} ba & \text{if} \quad n_a \leqslant 6 \\ c & \text{if} \quad n_a > 6 \end{cases} \tag{4-27}$$

and, using the notations of equations (4-17) and (4-18):

$$0_dR_\alpha = 5R_\alpha \qquad \text{i.e.} \qquad 0_dR(d) = \begin{cases} f(d) & \text{if} \quad n_d \leqslant 5 \\ f & \text{if} \quad n_d > 5 \end{cases} \tag{4-28}$$

$$0_eR_{-\alpha} = 5R_{-\alpha} \qquad \text{i.e.} \qquad 0_eR(e) = \begin{cases} g[e] & \text{if} \quad n_e \leqslant 5 \\ g & \text{if} \quad n_e > 5 \end{cases} \tag{4-29}$$

In the latter equations the parentheses denote a change of direction towards the left and the square brackets denote a change of direction towards the right (see Paragraph 2.1.3).

The operation sequence is thus:

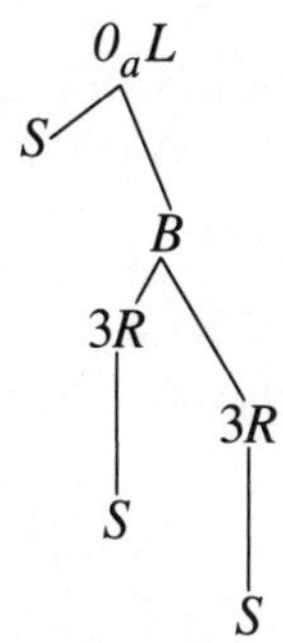

$$\begin{array}{l} a \rightarrow ba \quad \text{then} \quad c \\ b \rightarrow b \\ c \rightarrow de \\ d \rightarrow f(d) \quad \text{then} \quad f \\ e \rightarrow g(e) \quad \text{then} \quad g \\ f \rightarrow f \\ g \rightarrow g \end{array} \tag{4-30}$$

The first developmental stages are characterized by the activity of the a meristem (see Paragraph 2.1.1-C) until $n_a = 6$. During these first stages DS_{EE} behaves as the simple developmental system DS_A (eq. 1-1):

$$DS_{EE}(0) = a \qquad DS_{EE}(1) = ba \qquad \ldots \qquad DS_{EE}(3) = bbbbbba$$

Then the 0_aL becomes a T operation (eq. 4-27)

$$DS_{EE}(7) = bbbbbbc$$

and two symmetric subsystems are generated, the meristems of which are d and e respectively:

$$DS_{EE}(8) = bbbbbbde \qquad DS_{EE}(9) = bbbbbbf(d)g[e] \qquad \ldots$$

$$DS_{EE}(14) = bbbbbbf(f(f(f(f(f)))))g[g[g[g[g[g]]]]]$$

$$DS_{EE}(k) = DS_{EE}(14) \qquad \forall\, k > 14$$

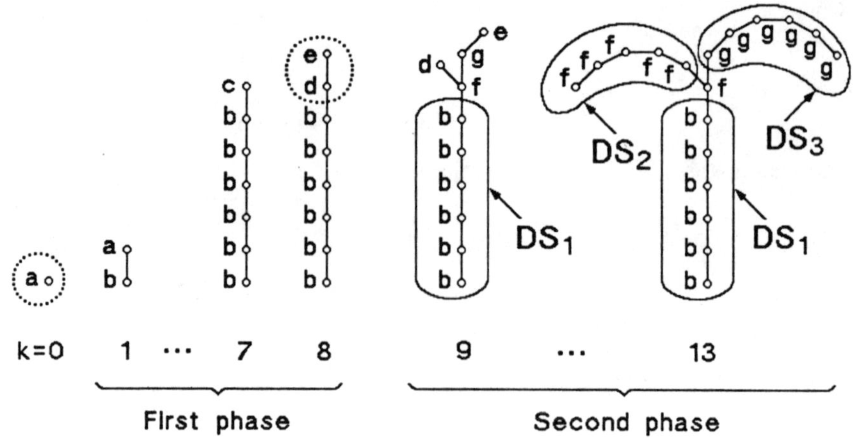

Fig. 4-12. Two-level development of system DS_{EE} (eqs. 4-26, 4-30).

In other words (see Figure 4-12):

1) During the first phase (k = 1 to 7) a subsystem DS_1 develops which consists of a rectilinear "tissue" of b cells (as in Figures 1-2 and 2-15).

2) During the second phase (k = 8 to 14) two symmetrical subsystems DS_2, DS_3 develop; they consist of spiral "tissues" of f and g cells respectively. The DS_1 subsystem plays the part of a passive "carrier".

4.5.2 Example

A) Analysis

These considerations will now be illustrated by a developmental process that results in Figure 4-13, which can be considered as the first approximation of the shape of an "adult" maple leaf (Fig. 4-14). It consists of six patterns which are assumed, for simplicity's sake, to be symmetric.

1) Pattern 1 can be generated by

$$DS_{FF} = \frac{(10L)\,S}{a,b} \tag{4-31}$$

which has been adapted from DS_A (eq. 1-1) by limiting to 10 the number of times the L operation is performed:

$$10L(a) = \begin{cases} ba & \text{if} \quad n_a \leqslant 10 \\ a & \text{if} \quad n_a > 10 \end{cases}$$

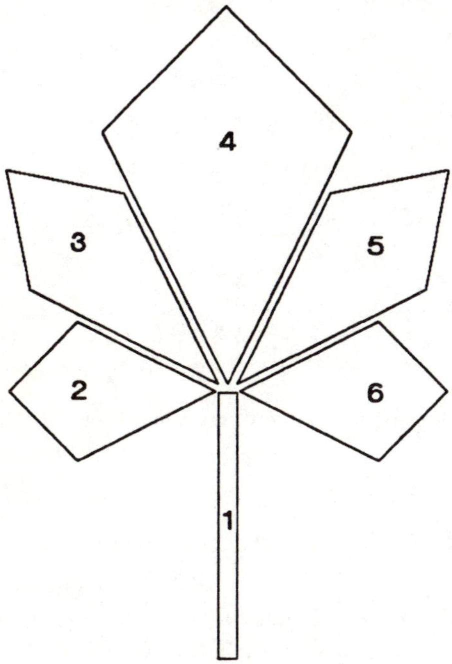

Fig. 4-13. Breaking down a maple leaf into six patterns.

Fig. 4-14. A maple leaf.

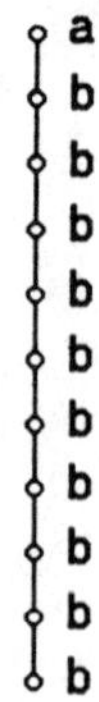

Fig. 4-15. $k = 10$ and 11 developmental stages of $D_{FF} = \dfrac{(10L)\,S}{a,b}$

Thus (Fig. 4-15)

$$DS_{FF}(10) = bbbbbbbbbba \tag{4-32}$$

2) Patterns 2 to 6 can be generated by the developmental system DS_H of Paragraph 2.1.1-C. It is seen in Figure 4-16 that its $k = 7$, 11 and 16 developmental stages provide satisfactory approximations respectively for the leaf patterns: 2 and 6; 3 and 5; 4.

B) SYNTHESIS

The six-pattern approximation shown in Figure 4-13 can be obtained as the result of a three-level development.

1) First phase: pattern 1 is obtained exactly as the DS_1 subsystem of Paragraph 4.5.1. It consists of the k = 11 developmental stage of DS_{FF}, the latter system being modified by replacing $10L$ by O_aL

$$O_aL(a) = \begin{cases} ba & \text{if} \quad n_a \leq 10 \\ c & \text{if} \quad n_a > 10 \end{cases} \tag{4-33}$$

(see eq. 4-27). Thus

$$DS'_{FF}(11) = bbbbbbbbbbc \tag{4-34}$$

2) Second phase: the terminal cell c divides into five cells, each of which will later generate one of the remaining patterns (2 to 6).

This can be implemented by having c be the initial condition of the developmental system

$$DS_{GG} = \frac{B\,B\,C_{\pm 90}\,C_{-90}\,S\,S\,S\,S\,S}{c,e,f,g,d,h,i,j,k} \tag{4-35}$$

which is characterized by the operation sequence

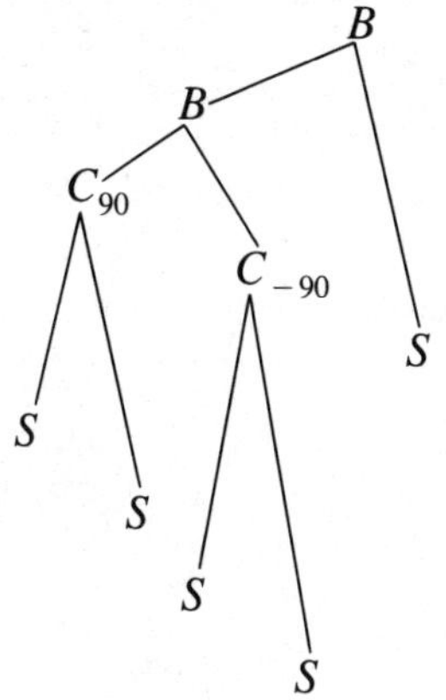

$$\begin{aligned} c &\to ed \\ e &\to fg \\ f &\to h(i) \\ g &\to j[k] \\ d &\to d \\ h &\to h \\ i &\to i \\ j &\to j \\ k &\to k \end{aligned} \tag{4-36}$$

whence (Fig. 4-17)

$$DS_{GG}(0) = c \qquad DS_{GG}(1) = ed \qquad DS_{GG}(2) = fgd$$
$$DS_{GG}(3) = h(i)j[k]d \tag{4-37}$$

3) Third phase: each of the five cells obtained gives rise to a developmental process of the type shown in Figure 4-16.

This can be implemented by DS_H systems (§ 2.1.1-C) modified by an operating system which "freezes" the development after the m-th stage, i.e. the L and C operations become stagnations after $k = m$:

$$DS_{Hm} = \frac{(O_aL)\,(O_bC)\,S\,(O_dL)\,S}{\text{a,b,c,d,e}} \tag{4-38}$$

where

$$\left.\begin{aligned} O_aL(a) &= \begin{cases} ba & k \leqslant m \\ a & k > m \end{cases} \\ O_bC(b) &= \begin{cases} b(c) & k \leqslant m \\ b & k > m \end{cases} \\ O_dL(d) &= \begin{cases} ed & k \leqslant m \\ d & k > m \end{cases} \end{aligned}\right\} \tag{4-39}$$

Thus (see Figure 4-18)

$$\text{patterns 2 and 6} = DS_{H7}(\mathrm{k}) \qquad \text{for} \qquad k \geqslant 7$$
$$\text{patterns 3 and 5} = DS_{H11}(\mathrm{k}) \qquad \text{for} \qquad k \geqslant 11$$
$$\text{pattern 4} = DS_{H16}(k) \qquad \text{for} \qquad k \geqslant 16$$

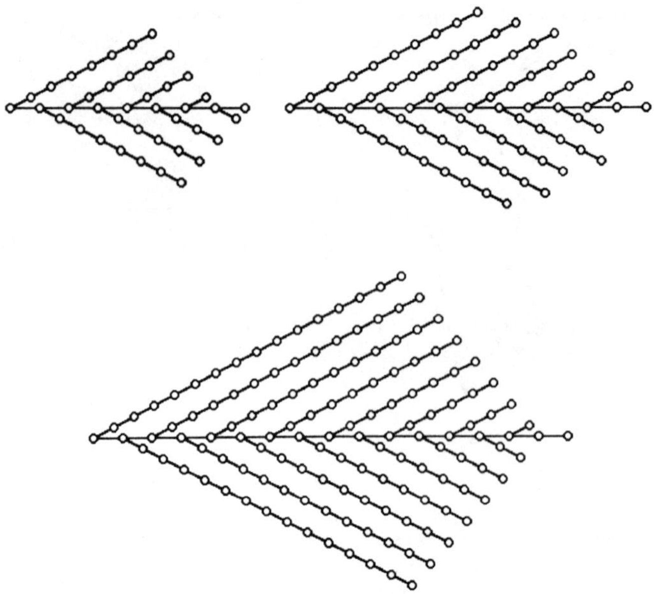

Fig. 4-16. k = 7, 11 and 16 developmental stages of $DS_H = \dfrac{\text{L C S L S}}{a,b,c,d,e}$

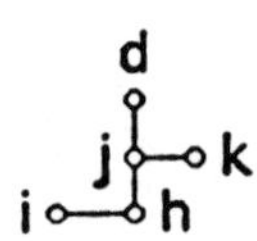

Fig. 4-17. Division of c cell into five cells (eqs. 4-36, 4-37).

The global generating word is finally

$$GW_{HH} = (0_aL) \quad \begin{array}{cccccccc} B & & B & & & C & & C \\ | & & \diagdown & & \diagup & \diagdown & \diagup & \diagdown \\ GW_{H16} & & & GW_{H7} & GW_{H7} & & GW_{H11} & & GW_{H11} \end{array} \tag{4-40}$$

The k-th developmental stage ($k \geqslant 16$) is shown in Figure 4-19.

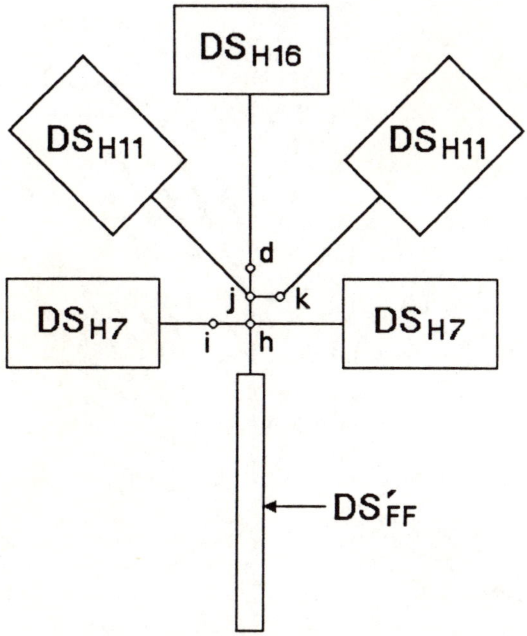

Fig. 4-18. Obtaining the configuration shown in Figure 4-13 with DS_{FF} (eq. 4-33) and DS_{Hm} (eq. 4-38). See figure 4-19.

4.6 ON GLOBAL OPERATING SYSTEMS

The operating systems considered so far consist of modifications of the elementary operations of the generating word

$$GW = \mathop{\mathrm{SEQ}}_{i=1}^{n} 0_i A_i \tag{4-4}$$

Now, the elements of a developmental system are grouped in subsystems in the same manner as the cells of a living organism are organized in tissues and organs (see Paragraph 2.2.2). This suggests the possibility of a *global operating system* which controls all the operations of a subsystem

$$GW = 0_{g1} (\mathop{\mathrm{SEQ}}_{i=1}^{m} A_i)\, 0_{g2} (\mathop{\mathrm{SEQ}}_{i=m+1}^{p} A_i)\, 0_{g3} (\mathop{\mathrm{SEQ}}_{i=p+1}^{n} A_i) \tag{4-41}$$

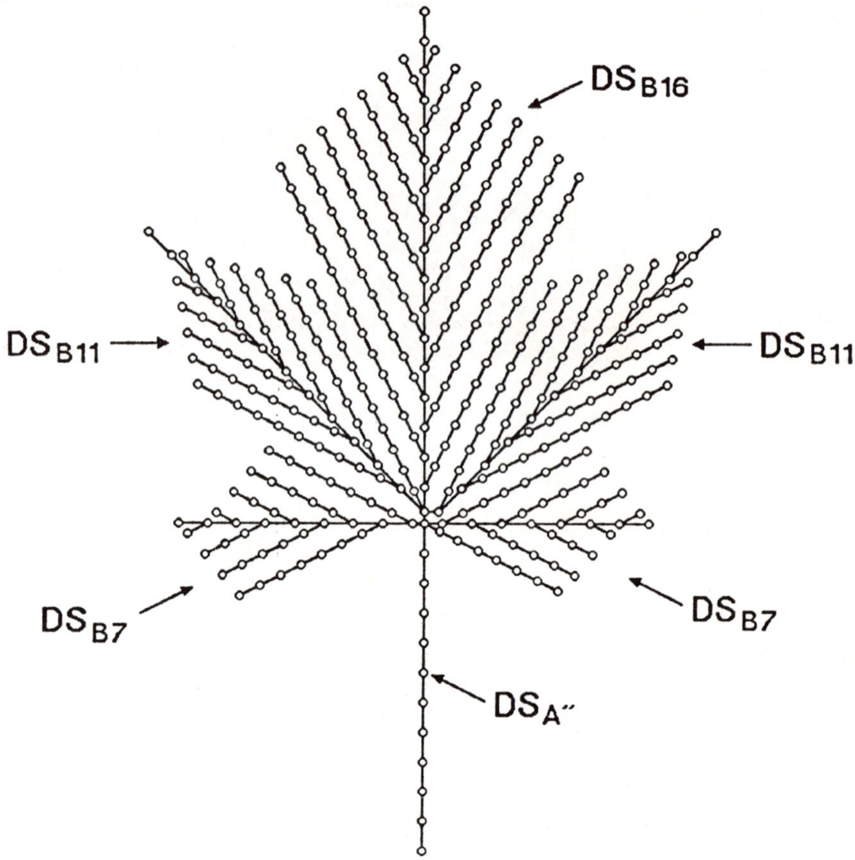

Fig. 4-19. $k = 16$ developmental stage of system DS_{HH} (eq. 4-40).

or of the whole system

$$GW = 0_g \,(\underset{i=1}{\overset{n}{\mathrm{SEQ}}}\, A_i) \tag{4-42}$$

Such global controls finally involve modifications of the individual A_i operations. Hence (4-42) can be broken down into

$$0_g \,(\underset{i=1}{\overset{n}{\mathrm{SEQ}}}\, A_i) = \underset{i=1}{\overset{n}{\mathrm{SEQ}}}\, 0_{gi} A_i \tag{4-43}$$

An example was given in Paragraph 4.5.2: the global development of

$$DS_H = \frac{LCSLS}{a,b,c,d,e} \tag{2-7}$$

is "frozen" after $k = m$ if the L and C operations are modified according to (4-39). In other words

$$0_g(LCSLS) = (0_aL)\,(0_bC)\,S\,(0_dL)S \tag{4-44}$$

A generating word of the (4-41) or (4-42) type can thus be expressed as a sequence of control symbols 0_{gi} and of operation symbols A_i without any parenthesis, just as an instruction to a computer. (Recall that the parentheses on the right-hand side of (4-44) have been merely inserted for clarity and have no semantic signification: see Section 4.1.)

4.7 CONCLUSION

The proposed model, despite its simplicity, accounts for many properties which developmental systems have in common with living organisms. In its original form (the generating word consists of letters B, C, L, R, T, S) it accounts for the properties of growth in size, of internal hierarchy, of regeneration after partial destruction, of mutation and for the possibility of cloning or grafting a system on another. If an operating system is introduced into the generating word, other properties can be accounted for, such as decay and dwindling, regrowth, "reproduction", sensitivity to external and to internal context, multilevel developement.

These facts are a consequence of the conformity of our axioms with some fundamental assumptions of cellular biology, the generating word of our model being the analog of the genetic code present in each cell of an organism.

We hope that with more sophisticated models, which we are now trying to develop, it will be possible to account for more complex properties of living organisms — and that our models will some day suggest new ideas and hypotheses to biologists.

REFERENCES

M.J. BERRIDGE (1985), Les molécules de communication dans la cellule, *Pour la science*, n. 98, pp. 134-146.

J.C. GILLE, S. WĘGRZYN and P. VIDAL (1988), Les phénomènes de repousse et de semailles dans les systèmes évolutifs de mot génétique circulaire, *Bulletin of the Polish Academy of Sciences (Technical Sciences)*, vol. 36, pp. 291-296.

J.C. GILLE, S. WĘGRZYN and P. VIDAL (1988), On some models for development systems. Part IX: generalized generating word and genetic code, *International Journal of Systems Science*, vol. 19, pp. 845-855.

J.C. GILLE, S. WĘGRZYN and P. VIDAL (1989), On some models for developmental systems. Part X: growth, decay and regrowth of developmental systems, *International Journal of Systems Science*, vol. 20, pp. 97-105.

J.L. POPOT (1987), La structure des protéines membranaires, *La Recherche*, vol. 18, pp. 1170-1181.

D.T. SUZUKI, A.J.F. GRIFFITHS and R.C. LEWONTIN (1981), *An Introduction to Genetic Analysis*, 2nd ed., Freeman, San Francisco, 919 p.

P. VIDAL, J.C. GILLE and S. WĘGRZYN (1987), Mot génétique de système évolutif et programme d'ADN, *Bulletin of the Polish Academy of Sciences (Technical Sciences)*, vol. 35, pp. 61-70.

P. VIDAL, J.C. GILLE and S. WĘGRZYN (1988), On some properties of developmental systems and their modelling, *Systems Analysis, Modelling, Simulation*, vol. 6, pp. 279-291.

P. VIDAL, S. WĘGRZYN and J.C. GILLE (1988), Accroissement et dépérissement des systèmes évolutifs à mot génétique linéaire, *Bulletin of the Polish Academy of Sciences*, vol. 36, pp. 285-290.

P. VIDAL, S. WĘGRZYN and J.C. GILLE (1989), On some models for developmental systems. Part XI: communication between the elements of a developmental system, *International Journal of Systems Science*, vol. 20, pp. 1795-1800.

J.D. WATSON (1976), *Molecular Biology of the Gene*, 3rd. ed., Benjamin, Menlo Park, 739 p.

S. WĘGRZYN, P. VIDAL and J.C. GILLE (1990), On some models for developmental systems. Part XII: More on internal context sensitivity, *International Journal of Systems Science*, vol. 21 [to appear].

OTHER BOOKS BY THE SAME AUTHORS

Genetyka procesów rozwoju [The Genetics of Developmental Processes], Polish version of this monograph (Polytechnic of Silesia Press, Gliwice, 1988, 70 p.).

In collaboration with O. Palusiński:

Introduction à l'étude de la stabilité dans les espaces métriques (Dunod, Paris, 1971, 73 p.). First published in Polish (1970).

Books by S. Węgrzyn:

Calcul opérationnel en électrotechnique (Gauthier-Villars, Paris, 1967, 295 p.). First published in Polish (1955; 2nd ed. 1960). Also in Slovak (1958) and in German (1962).

Przebiegi nieustalone w elektrycznych liniach i układach łańcuchowych [Transients in electrical lines and chain networks] (Państwowe Wydawnictwo Naukowe, Warsaw, 1958, 140 p.).

Podstawy informatyki [Fundamentals of Computer Science] (Państwowe Wydawnictwo Naukowe, Warsaw, 1982, 394 p.).

In collaboration with A. Bukowy and Z. Pogoda:

Les Bases de l'automatique industrielle (Dunod, Paris, 1965, 328 p.). First published in Polish (1963; 5th ed. 1980).

Books by J.C. Gille:

Introduction aux systèmes asservis non linéaires (Dunod, Paris, 1977; 2nd ed. 1984, 126 p.).

In collaboration with P. Decaulne and M. Pélegrin:

Feedback Control Systems. Analysis, Synthesis, and Design (McGraw-Hill, New York, 1959, 793 p.). First published in French (1956). Also in Polish (1961) and in Russian (1961).

Théorie et Calcul des asservissements (Dunod, Paris, 1958; 3rd ed. 1963, 321 p.). Also in German (1960, 3rd ed. 1968), in Rumanian (1963), in Italian (1966) and in Spanish (1967, 2nd ed. 1971).

Les Organes des systèmes asservis (Dunod, Paris, 1959; 3rd ed., 1965, 463 p.). Also in German (1962; 2nd ed. 1967) and in Rumanian (1963).

Dynamique de la commande linéaire (Dunod, Paris, 1967; 8th ed. 1988, 524 p.)

Théorie et Calcul des asservissements linéaires (Dunod, Paris, 1967; 9th ed. 1990, 489 p.).

Systèmes asservis non linéaires (Dunod, Paris, 1967; 5th ed., 1987, 3 vol., 163 + 151 + 219 p.)

Introduction aux systèmes asservis extrémaux et adaptatifs (Dunod, Paris, 1976, 92 p.).

In collaboration with P. Decaulne, M. Carpentier and M. Pélegrin:

Problèmes d'asservissements avec solutions (Dunod, Paris, 1959; 4th ed. 1971, 256 p.). Also in Polish (1961) and in German (1963; 2nd ed. 1967).

.../.

In collaboration with M. Clique:

La Représentation d'état pour l'étude des systèmes dynamiques (Eyrolles, Paris, 1975, 2 vol., 192 + 109 p.).

Calcul matriciel et Introduction à l'analyse fonctionnelle (Lidec, Montréal, 1979; 4th ed. 1989, 3 vol., 163 + 116 + 124 p.). Also in Polish (1977; 2nd ed. 1986, 320 p.).

Systèmes linéaires. Équations d'état (Eyrolles, Paris, 1984; 2nd ed. 1990, 203 p.).

Calcul matriciel. Exercices et problèmes (Lidec, Montréal, 1984; 2nd ed. 1988, 223 p.)

Books by P. Vidal:

Non-Linear Sampled-Data Systems (Gordon and Breach, New York, 1967, 346 p.). Also in French (1968) and in Russian (1974).

Non-Linear Sampled-Data Systems. Exercizes and Problems (Gordon and Breach, New York, 1972, 103 p.). First published in French (1970).

Aide-Mémoire d'automatique (Dunod, Paris, 1978; 2nd ed. 1985, 196 p.).

Editor of the series **Automatique** and **Robotique** (Techniques de l'ingénieur, Paris, since 1982).